HANDBOOK FOR ESTIMATING PHYSICOCHEMICAL PROPERTIES OF ORGANIC COMPOUNDS

MARTIN REINHARD

Department of Civil and Environmental Engineering, Stanford University, Stanford, California, USA

AXEL DREFAHL

Institute for Physical Chemistry, Technical University, Bergakademie Freiberg, Freiberg, Sachsen, Germany

A WILEY-INTERSCIENCE PUBLICATION

JOHN WILEY & SONS, INC.

New York • Chichester • Weinheim • Brisbane • Singapore • Toronto

Copyright © 1999 by John Wiley and Sons, Inc. All rights reserved.

Published simultaneously in Canada.

Library of Congress Cataloging-in-Publication Data

Reinhard, Martin.
    Handbook for estimating physicochemical properties of organic
      compounds / Martin Reinhard and Axel Drefahl.
        p.   cm.
    "A Wiley-Interscience publication."
    Includes bibliographical references and index.
    ISBN 0-471-17264-2 (cloth : alk. paper)
    1. Organic compounds—Handbooks I. Drefahl, Axel,
    II. Title.
QD257.7.R45   1998
547—dc21   98-15969
                                      CIP

Printed in the United States of America.

10 9 8 7 6 5 4 3 2 1

CONTENTS

PREFACE

The purpose of this *Handbook* is to introduce the reader to the concept of property estimation and to summarize property estimation methods used for important physicochemical properties. The number of estimation methods available in the literature is large and rapidly expanding. This book covers a subset judged to have relatively broad applicability and high practical value. Property estimation may involve the selection of an appropriate mathematical relationship, identification of similar compounds, retrieval of data and empirical constants, standard adjustments for nonpressure temperature, and examination of original literature. To facilitate this often tedious task, we have developed the "Toolkit for Estimating Physicochemical Properties" (Reinhard and Drefahl, 1998), hereafter referred to as the Toolkit.

In some cases, property estimation methods may yield results that are nearly as good as measured values. However, estimates often deviate from the accurate value by a factor of 2 or more and may be considered order-of-magnitude estimates. For many applications, such estimates are adequate. Some of the estimation methods discussed are qualitative rules that indicate that a property of the query is greater or smaller than a given value. Generally, the accuracy of property estimation methods is difficult to assess and has to be discussed on a case-by-case basis. Chemical intuition remains an important element in all property estimations, however.

ACKNOWLEDGMENTS

We are indebted to Jeremy Kolenbrander for reviewing the book and thank him and Frank Hiersekorn for contributing to DESOC, the precursor to the Toolkit. Tilman Kispersky and Katharina Glaser helped to prepare the bibliography. Funding for this project was provided in part by the Office of Research and Development, U.S. Environmental Protection Agency, under Agreement R-815738-01 through the Western Region Hazardous Substance Research Center, and by Aquateam, Oslo, Norway. The content of the book does not necessarily reflect the view of these organizations.

MARTIN REINHARD

Stanford University

AXEL DREFAHL

Institute for Physical Chemistry

CHAPTER 1

OVERVIEW OF PROPERTY
ESTIMATION METHODS

1.1 INTRODUCTION

Purpose and Scope Knowing the physicochemical properties of organic chemicals is a prerequisite for many tasks met by chemical engineers and scientists. An example of such a task includes predicting a chemical's bioactivity, bioavailability, behavior in chemical separation, and distribution between environmental compartments. Typical compounds of concern include bioactive compounds (biocides, drugs), industrial chemicals and by-products, and contaminants in natural waters and the atmosphere. Unfortunately, there are very limited or no experimental data available for most of the thousands of organic compounds that are produced and often released into the environment. In the United States, the Toxic Substances Control Act (TSCA) inventory has about 60,000 entries and the list is growing by 3000 every year. Some 3000 chemicals are submitted to the United States Environmental Protection Agency (EPA) for the premanufacture notification process, most completely without experimental data. The data for more than 700 chemicals on the Superfund list of hazardous substances are limited [1]. For the many compounds without experimental data, the only alternative to making actual measurements is to approximate values using estimation methods. Estimated values may be sufficiently accurate for ranking compounds with respect to relevant properties. Such rankings for example, allow investigators qualitatively prediction of compound behavior in environmental systems during waste treatment, chemical analysis, or bioavailability.

The purpose of this handbook is to introduce the reader to the concept of property estimation and to summarize property estimation methods used for some important physicochemical properties. The number of estimation methods available in the literature is large and rapidly expanding and this book covers only a subset. The methods that were selected for discussion were judged to have relatively broad applicability and high practical value. Property estimation methods that yield results better than approximately 20% are termed *quantitative*. However, estimates often may deviate from the accurate value by a factor of 2 and the estimate may be considered *semiquantitative*. An example of a semiquantitative property–property estimation method is that for the octanol/water partition coefficient, K_{ow}. Estimates for log K_{ow} typically deviate by a factor of 2 or more. Some of the methods discussed are qualitative rules that indicate that a property of the query is greater or smaller than a given value or provide an order-of-magnitude estimate.

Classes of Estimation Methods   Table 1.1.1 summarizes the property estimation methods considered in this book. Quantitative property–property relationships (QPPRs) are defined as mathematical relationships that relate the query property to one or several properties. QPPRs are derived theoretically using physicochemical principles or empirically using experimental data and statistical techniques. By contrast, quantitative structure–property relationships (QSPRs) relate the molecular structure to numerical values indicating physicochemical properties. Since the molecular structure is an inherently qualitative attribute, structural information has first to be expressed as a numerical values, termed *molecular descriptors* or *indicators* before correlations can be evaluated. Molecular descriptors are derived from the compound structure (i.e., the molecular graph), using structural information, fundamental or empirical physicochemical constants and relationships, and stereochemcial principles. The molecular mass is an example of a molecular descriptor. It is derived from the molecular structure and the atomic masses of the atoms contained in the molecule. An important chemical principle involved in property estimation is *structural similarity*. The fundamental notion is that the property of a compound depends on its structure and that similar chemical stuctures (similarity appropriately defined) behave similarly in similar environments.

TABLE 1.1.1   Classes of Property Estimation Methods

Method	Predictor Variable
Quantitative property–property relationships (QPPRs)	Property
Quantitative structure–property relationships (QSPRs)	Molecular descriptor
Group contribution models (GCMs)	Fragment constants
Similarity-based models	
Between isomeric compounds	Molecular descriptor
Between homologous compounds	Fragment constant for CH_2
Between similar compounds	
Group interchange models (GIMs)	Properties of similar compound(s), fragment constants
Nearest-neighbor models	Properties of k similar compounds
Mixed models	Combinations of the above

Properties are physicochemical or biological characteristics of compounds that can be expressed qualitatively or quantitatively. Most physicochemical properties generally are related to and depend on one another in some ways and to varying degrees. Table 1.1.2 summarizes the properties that are considered in this book.

Chemists are trained to recognize the significance of compound similarity and dissimilarity in the context of the problem at hand. This "cognitive" approach, when

TABLE 1.1.2 Summary of Properties

Property*	Symbol
Density	ρ
Molar volume	V_M
Refractive index	n
Molar refraction	R
Surface tension	s
Parachor	—
Viscosity	η
Vapor pressure	p_v
Enthalpy of vaporization	ΔH_v
Boiling point	T_b
Melting point	T_m
Aqueous solubility	S_w
$K_{\text{air-water}}$	K_{aw}
$K_{\text{octanol-water}}$	K_{ow}
$K_{\text{organic carbon-water}}$	K_{oc}

*Note: All properties indicated can be estimated using the Toolkit.

done by humans rather than by computers, is usually slow and limited to a small set of compounds. Moreover, it lacks quantitative rigor. Computerized algorithms have made it possible rapidly to quantify the structural similarity of thousands of compounds, to recognize the structural differences, and to evaluate the relationships between structure and properties. Several algorithms have been developed to translate molecular graphs into a computer readable language suitable for the evaluation of chemical structures, such as the determination of chemical structure similarity. Definitions of the basic concepts, descriptions, and references for further study are discussed below. Understanding of these principles will be helpful when using computer-aided property estimation techniques and assessing the validity of results.

Chemical property estimation is the process of deriving an unknown property for a query compound from available properties, molecular descriptors, or reference compounds. The selected subset of the reference compounds depends on the query and is termed a *training set*. Training sets may consist of narrowly defined classes of closely related compounds such as structural isomers and homologous compounds. Figure 1.1.1 provides an overview of the data needs and the information flow in four property estimation approaches. To illustrate these examples, benzene and toluene are considered a subset of a larger data set with n measured compounds and chlorobenzene is the query compound. The n compounds with known octanol–water partition coefficients, K_{ow}, represent the training set. From the K_{ow} data set and the water solubility, S_w data set can be derived the property/property relationship that relates S_w to K_{ow}. The compounds used as specific examples, benzene, toluene, and chlorobenzene, are similar to each other in that they are all hydrophobic and of relatively low molecular weight. Furthermore, solubilization in water is a process similar to partitioning in octanol–water in that the solute distributes itself between a polar phase (water) and an apolar phase in both cases. The relationship between K_{ow} and S_w relates two different properties and is called a quantitative property/property relationship (QPPR). In the example shown, the QPPR is

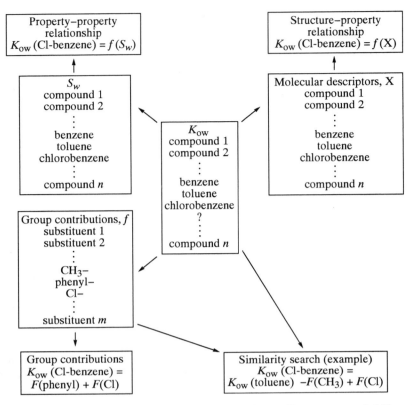

Figure 1.1.1 Examples of property estimation techniques (S_w = water solubility; K_{ow} = octanol-water partition coefficient). Chlorobenzene is the query compound. F are fragment or atom constants; f is a property-property or a structure-property relationship.

used to estimate the K_{ow} of the query compound chlorobenzene. Similarly, a training set can be used to develop a structure/property relationship by evaluating the relationship between molecular descriptors and a property. The example shown in Figure 1.1.1 uses a training set of K_{ow} data to establish a relationship between K_{ow} and the molecular descriptor X. Such relationships are called quantitative structure / property relationships (QSPR). This QSPR can then be used to estimate the K_{ow} of a query. Of course, to obtain statistically meaningful results, the training set must contain a minimum number of entries and the properties of the compounds represented must span an adequate property range. For a few isomeric groups and homologous series, rules have been derived that allow to predict the effect of structural modification on a compound property [Sections 1.2 and 1.3]. Generally, QSPR and QPPR methods are limited to compounds and properties falling within the range given by the training set used to develop the particular relationship [Sections 1.4 and 1.5].

Another frequently used method to derive empirical relationships between structure and property is to divide the structure into chemically logic parts such as groups of atoms (functional groups) and to assign each group a contribution to the property of the whole molecule. This approach is termed the group contribution model (GCM). Since groups cannot be measured individually, it is necessary to derive group contributions by comparing the properties of compounds containing the individual groups as part of a molecule and to statistically evaluate the contributions of each group [Section 1.6]. In the example shown in Figure 1.1.1, the K_{ow} is obtained as the sum of two group contributions, those of the phenyl group and the chlorine atom.

Similarity-based approaches are based on the assumption that closely related compounds have closely related properties. These approaches use as a starting point one or several, k, closely related compounds (the k nearest neighbors, kNN) with known properties. Then some model, such as averaging or a group contribution model, is used to further approximate the property value of the query. Obviously, the closer the relationship with the query the better the final result will be. Traditionally, the kNN approach has been used in categorical or semiquantitative property estimation. In the example shown in Figure 1.1.1, toluene has been identified as a compound similar to chlorobenzene. The K_{ow} of chlorobenzene is then obtained by subtracting the group contribution of the methyl group, $f(CH_3)$ and adding the group contribution of Cl, $f(Cl)$. Many other approaches are possible, and the development of kNN approaches are subject of current research.

Often, it is important to know not only the property itself at a standard temperature but also its temperature dependence. Temperature functions are available for a wealth of fluid compounds, such as solvents. However, these functions are compound specific. For limited sets of compounds, functions have been developed that describe properties as a function of both molecular structure and temperature (Section 1.9).

Computer-Aided Property Estimation Computer-aided structure estimation requires the structure of the chemical compounds to be encoded in a computer-readable language. Computers most efficiently process linear strings of data, and hence linear notation systems were developed for chemical structure representation. Several such systems have been described in the literature. SMILES, the Simplified Molecular Input Line Entry System, by Weininger and collaborators [2–4], has found wide acceptance and is being used in the Toolkit. Here, only a brief summary of SMILES rules is given. A more detailed description, together with a tutorial and examples, is given in Appendix A.

SMILES is based on the "natural" grammer of atomic symbols and symbols for bonds. The most important rules are as follows:

1. Atoms are represented by their atomic symbols, (e.g., B, C, N, O, P, S, F, Cl, Br, I). Hydrogen atoms are usually omitted.
2. Atoms in aromatic rings are specified by lowercase letters. For example, the nitrogen in an amino acid is represented as N, the nitrogen in pyridin by n, and carbon in benzene by c.
3. Single, double, triple, and aromatic bonds are represented by the symbols $-$, $=$, #, and : , respectively. Single and aromatic bonds may be omitted.
4. Branches are represented by enclosure in parentheses.

These rules are illustrated by the examples in Table 1.1.3. For most structures several SMILES can be deduced, depending on the starting point. All SMILES are

TABLE 1.1.3 Examples of SMILES Notations

Compound Name	Formula	SMILES	Comment
Methane	CH_4	C	H atoms suppressed
Methylamine	CH_3NH_2	CN	Single bond suppressed
Hydrogen Cyanide	HCN	C#N	Triple bond not suppressed
Vinyl chloride	$ClHC=CH_2$	ClC=C	Double bond not suppressed
Isobutyric acid	$CH_3CH(CH_3)COOH$	CC(C)C(=O)O	Parenthesis indicate branching
Benzene	C_6H_6	c1ccccc1	Aromatic bonds omitted, ring closure at numbers following c
t-Butylbenzene	$C_6H_5C(CH_3)_3$	c1ccccc1C(C)(C)C	Branching groups indicated by parentheses

valid. A computer algorithm can be used to identify the unique SMILES notation that is actually used for computer processing [3] (see Appendix A).

1.2   RELATIONSHIPS BETWEEN ISOMERIC COMPOUNDS

Two molecules share an isomeric relationship if they have the same molecular formula. All molecules with the same molecular formula constitute a set of structural isomers and are to some degree similar. However, they may have different chemical constitutions, as indicated in Figure 1.2.1 for 1-butanol and five structural isomers. Any two of these molecules placed in the same row make a pair of constitutional isomers. For the purpose of property estimation, it is helpful to further classify the constitutional isomers according to type and position of the functional groups and branching of the isomers. In the dicussion that follows, we focus on two different types of isomeric sets: positional isomers and branched isomers.

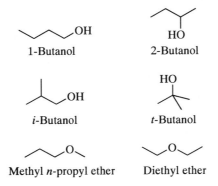

1-Butanol   2-Butanol

i-Butanol   t-Butanol

Methyl n-propyl ether   Diethyl ether

Figure 1.2.1   Six possible isomers with the molecular formula $C_4H_{10}O$.

Positional Isomers   *Positional isomers* differ in the position where a functional group occurs in a molecule. In Figure 1.2.1, 1-butanol and 2-butanol are positional isomers with the position of the hydroxyl group indicated by the prefixes 1 and 2, respectively. Similarly, methyl *n*-propyl ether and diethyl ether are positional isomers, as reflected in their synonym names 2-oxapentane and 3-oxapentane, with the prefixes 2 and 3 indicating the position of the ether group, respectively.

Branched Isomers   *Branched isomers* differ in the degree of branching of their alkyl groups. 1-Butanol, *i*-butanol, and *t*-butanol are branched isomers (including the unbranched 1-butanol for the sake of completeness) with increasing degree of branching in their alkyl group. The unbranched isomer is often denoted as a *normal* isomer. Besides the atoms of the functional group, the normal isomer consists solely of primary and secondary C atoms, corresponding to methyl and methylene groups, respectively. In contrast, branched isomers contain tertiary and/or quaternary carbon atoms.

Properties of Isomers   By definition, isomers have equal molar masses. Many properties correlate significantly with the molar mass. It follows, then, that properties of isomeric compounds in such a class should be approximately equal. However, such generalizations should be applied with great caution. For example, anthracene and phenanthrene are constitutional isomers but have aqueous solubilities differing by a factor of about 100 [5]. In certain cases the properties for a set of isomers are well presented in terms of a property interval and a mean isomer value, as has been done for tetrachlorobenzyltoluenes (TCBT). TCBTs constitute a class of positional isomers with 96 possible congeners. The general structure of TCBT is indicated in Figure 1.2.2. For nine TCBs, log K_{ow} values have been measured at 25°C ranging from 6.725 ± 0.356 to 7.538 ± 0.089 with a mean isomer value of 7.265 ± 0.244 [6].

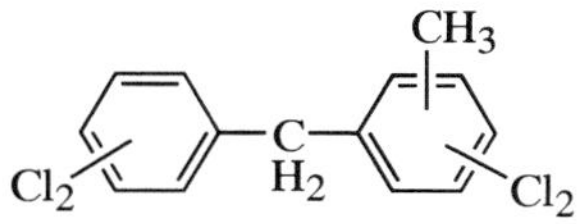

Figure 1.2.2   Generalized structure of tetrachlorobenzytoluene isomers. One ring is substituted for by two chlorine atoms and one ring by two chlorine atoms and a methyl group.

Stereoisomers   Structural isomers having an identical chemical constitution but exhibiting differences in the spatial arrangement of their atoms are called *stereoisomers* [7]. One case of stereoisomerism, denoted *asymmetric chirality*, comprises molecules that are mirror images of each other. Such pairs of molecules are called *enantiomers*. Figure 1.2.3 illustrates the two chiral molecules of 1-bromo-1-chloroethane. The line in the middle represents a symmetry plane. Note that it is

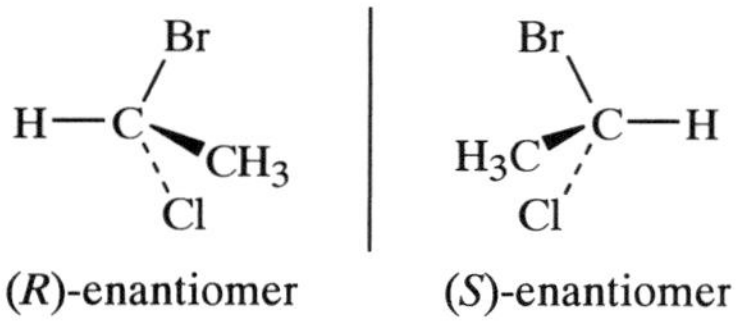

(*R*)-enantiomer   (*S*)-enantiomer

Figure 1.2.3   A pair of enantiomers shown image and mirror image.

cis-1,2-Difluoroethene   trans-1,2-Difluoroethene

Figure 1.2.4   A pair of diastereoisomers.

not possible to superimpose the two molecules by rotation and translocation. The two structures are related to each other as the left and right hands.

Stereoisomers that are not enantiomers are *diastereoisomers*. For example, *cis*- and *trans*-1,2- difluoroethene (Figure 1.2.4), constitute a pair of diastereoisomers.

Properties of Enantiomers   The spatial distances between atoms within an entiomer and the corresponding spatial distances between atoms within its enantiomeric counterpart are pairwise identical. Therefore, two enantiomers have equal energy contents [7] and will display identical molecular properties except in their interactions with other stereoisomers and light. The selective molecular recognition—by a receptor or biocatalyst, for example—allows the design of powerful separation techniques to detect enantiomers and to yield samples of high purity [8–10]. This specific interaction of stereoisomers has important biological and environmenal consequences. The effectiveness and toxicology of drugs depends on enantiomeric selectivity and purity. For example, the sedative thalidomide, prescribed to pregnant women as a racematic mixture, turned out to cause birth defects in children, whereas the pure R-enantiomer worked fine [11].

Properties of Diastereomers   In contrast to enantiomeric pairs, the correpond-ing spatial distances in diastereomeric pairs are not all identical. For example, *cis*- and *trans*-1,2-difluoroethene (Figure 1.2.4), differ in their F–F and H–H distances. This results into different energy contents and different properties between diastereomeric molecules. The difference in properties of diastereomers is illustrated with *cis*- and *trans*-1-phenyl-1,3-butadiene, which show markedly different physico-chemical properties [12] (Figure 1.2.5). Further investigation of stereochemical isomers is beyond the scope of this book, and discussion in subsequent chapters is limited to constitutional isomers.

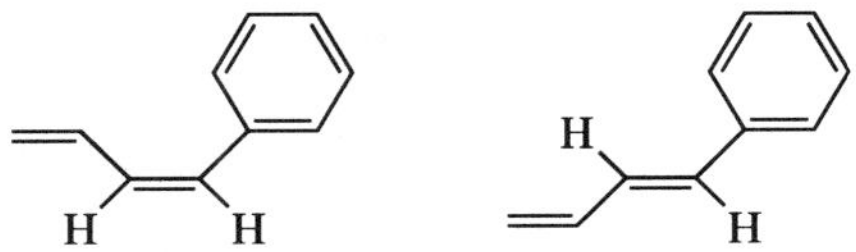

cis-1-Phenyl-1,3-butadiene
$T_m = (-56.99 \pm 0.04)\,°C$
$d_4^{25} = 0.9197$
$n_D^{25} = 1.5822$

trans-1-Phenyl-1,3-butadiene
$T_m = (4.52 \pm 0.04)\,°C$
$d_4^{25} = 0.9232$
$n_D^{25} = 1.6089$

Figure 1.2.5   Chemical structures of *cis*- and *trans*-1-phenyl-1,3-butadiene and their normal melting point, T_m, specific gravity, d_4^{25}, and the refractive index, n_D^{25}.

Structure–Property Relationships for Isomers   Structure–property relation-ships for isomers may indicate an increase or decrease in properties as a function of

(1) branching of the carbon skeleton or (2) the position of the substituents on the carbon skeleton. As an example, branching of alkyl groups tends to decrease the boiling point, T_b, of a compound. This observation can be stated as a qualitative rule:

$$T_b^p(n\text{-butyl}) < T_b^p(iso\text{-butyl}) < T_b^p(sec\text{-butyl}) < T_b^p(tert\text{-butyl}) \qquad \text{(R-1.2.1)}$$

where T_b^p is the boiling point at pressure p.

Structure–property rules in this book are presented in boxes along with an identifier of the form R-c.s.i, where c is the chapter number, s is the section number, and i is an index in that section. In some instances, similar structure–property relationships can be expressed quantitatively. In these cases, the difference in a property value, ΔP, for structural differences are indicated.

Number of Possible Isomers The number of isomers that may exist for a given molecular formula is known for special cases and it can be very large. For example, there are 262,144 (equal to 2^{18}) stereoisomers with the molecular formula of boromycin, $C_{45}H_{74}O_{15}BN$ [13]. A short historical introduction to the enumeration of isomeric acyclic structures has been given by Trinajstic' [14]. Coffman, Henze, and Blair have analyzed the numbers of possible alkene and alkyne isomers [15–17]. The interested reader is referred to an article [18] illustrating isomer counting of ter-, quater-, quinque-, and sexithienyls, compounds containing three, four, five and six thiophene rings, respectively. It is fun to do as an exercise and is useful in research on polythienyls as potential insecticides and as electrically conductive polymers.

1.3 RELATIONSHIPS BETWEEN HOMOLOGOUS COMPOUNDS

A set of homologous compounds consists of successive members differing in their molecular structure exactly by multiples of CH_2. Such a set is called a *homologous series*. Homologous compounds are similar in that they share the same basic carbon skeleton except for one or several inserted (or deleted) methylene groups. The incremental contribution of a methylene group to a property is small compared to the contribution of the parent, and rules that predict the properties of homologous compounds are based on compound similarity. The number of CH_2 groups is denoted as N_{CH_2}. Figure 1.3.1 shows, as an example, the first five members of the 1-iodoalkane series, along with their N_{CH_2} values and their molar mass, M.

The first compound of a series is called the *base member* and the following ones are called *derived members*. Note that in this definition N_{CH_2} does not account for the CH_2 group contained in the methyl group, CH_2–H. This definition is applied to be consistent with the group definition in most of the group contribution models (see Section 1.6), where H atoms are usually considered as parts of groups but not as groups by themselves. Thus, to avoid "isolated H atoms," treatment of methyl groups as a whole is recommended.

The molar mass increment for CH_2 is $14.027 \text{ g mol}^{-1}$. The following relation exists between any homologous member and its base member:

$$M(\text{derived member}) = (14.027 \text{ g mol}^{-1})N_{CH_2} + M(\text{base member}) \qquad (1.3.1)$$

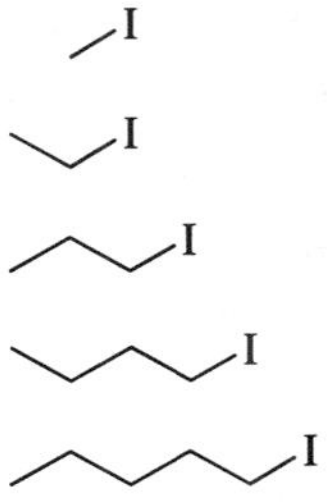

Iodomethane:	$\diagup{}^{I}$	$N_{CH_2} = 0$	$M = 141.94 \text{ g mol}^{-1}$
Iodoethane:	$\diagdown\diagup{}^{I}$	$N_{CH_2} = 1$	$M = 155.97 \text{ g mol}^{-1}$
1-Iodopropane:	$\diagup\diagdown\diagup{}^{I}$	$N_{CH_2} = 2$	$M = 169.99 \text{ g mol}^{-1}$
1-Iodobutane:	$\diagdown\diagup\diagdown\diagup{}^{I}$	$N_{CH_2} = 3$	$M = 184.02 \text{ g mol}^{-1}$
1-Iodopentane:	$\diagup\diagdown\diagup\diagdown\diagup{}^{I}$	$N_{CH_2} = 4$	$M = 198.05 \text{ g mol}^{-1}$

Figure 1.3.1   First five members of the 1-iodoalkane series together with their N_{CH_2} values and their molar mass, M.

This relation has been applied for other properties by substituting, for example, V_M or R_D for M and evaluating the corresponding coefficients. Usually, if such a property is known for three or more members of a homologous series, a relation can be derived for a given property, in analogy to eq. 1.3.1 by simple linear regression. The derived relationship, then, may be used to interpolate or extrapolate the property values to other homologous members. Such simple linear relationships for homologous series, however, are only approximations except for M. Therefore, other analytical functions have been studied to represent quantitative N_{CH_2}–property relationships. They have to be employed with caution when (1) phase changes are involved, and (2) the odd–even effect plays a role.

Phase Change   In many cases, homologous relationships are valid only for those member compounds that share the same physical state. Most common are relationships that apply to compounds in their liquid state.

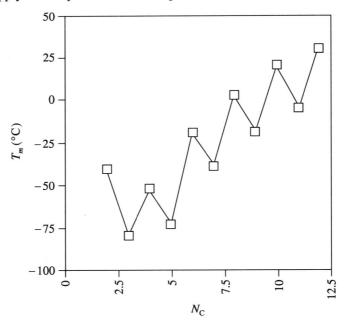

Figure 1.3.2 α,ω-Dimercaptans with T_m plotted against N_c. (*Source*: Reprinted with permission from Ref. 19. Copyright (1943) American Chemical Society.)

Odd–Even Effect The odd–even effect refers to the dependence of certain properties on the number of carbon atoms in a molecule, N_C. In such cases, properties of compounds containing a straight chain of CH_2 groups alternate with N_C. Typical examples of the odd–even effect can be found in diagrams that depict the melting points against N_C. An example is presented in Figure 1.3.2 showing the graphs of T_m against N_C for α,ω-dimercaptans [19]. Similar graphs have been published for such diverse series as alkanoic acids and their anhydrides [20], alkyl alkanoates [21], alkyl *p*-nitrobenzoates [22], and mono- and dialkyl ethers of stilboestrol [23]. Burrows [24] presents examples of the odd–even effect for properties other than the melting point temperature, including transition point properties and solubility behavior. In addition, he discusses the odd–even effect with respect to the stereochemical configuration and packing properties of the alkyl chain in the solid phase.

1.4 QUANTITATIVE PROPERTY–PROPERTY RELATIONSHIPS

A quantitative property–property relationship (QPPR), is a function that relates a property Y to one or several (m) other properties, $P_1, P_2, \ldots, P_m$:

$$Y = f(P_1, P_2, \ldots, P_m) \tag{1.4.1}$$

QPPR can be derived from thermodynamic principles or by statistical analysis of measured data. In the latter case, a set of compounds for which Y and $P_1, P_2, \ldots, P_m$ are known is required to develop the model (the training set). An additional *evaluation set* of compounds with known $Y, P_1, P_2, \ldots, P_m$ is recommended to evaluate the reliability and predictive capability of the model proposed. For a detailed description of the statistical methods, the reader is referred to [25], standard statistical texts, and to articles listed in the Toolkit Bibliography.

Application of a specific QPPR consistent with eq. 1.4.1 to estimate Y for a query compound requires the following:

1. $P_1, P_2, \ldots, P_m$ are known for the query compound.
2. The query compound belongs to the same compound class(es) defined by the training and evaluation sets.

In addition, one has to qualify the estimation result by identifying further possible limitations of the used model. For example, if a model applies to liquids only, one has to assure that the query compound is a liquid.

In the example shown in Figure 1.1.1, the water solubilities and the octanol–water partition coefficients of benzene, chlorobenzene, and toluene are related directly through the QPPR $K_{ow} = f(S_w)$. In this case, only one property, the water solubility, is used as the predictor variable. Chlorobenzene, the query, is considered similar to toluene and benzene because it contains one aromatic ring. The chlorine substituent is hydrophobic and bulky, similar to the methyl group of toluene. If the range of compounds is expanded to n other compounds, the applicability of the QPPR is expanded to all compounds similar to the set of n compounds included in the training set.

The applicability of a model for estimating a given query should be considered carefully. This book has been written to support the user in the verification process. The reviewed models are described along with their application range. This range is usually given by the substance classes used to develop the model. If other limitations are significant, these are either stated or the reader is referred to original sources.

In the literature, QPPRs are represented with varying details about the model derivation process. Statistical parameters, training and evaluation set information, and specification of the applicability range differ from publication to publication. Although guidelines for the application of QPPRs and QSPRs have been proposed [26], they are not always followed consistently. In this book, QPPRs are presented in the following form:

$$Y = A_0 + A_1 P_1 + A_2 P_2 + \cdots + A_M P_M \quad n = \ldots, \quad s = \ldots, \quad r = \ldots, \quad F = \ldots$$
$$(\text{c.s.i.})$$

where the expression on the left presents the particular model equation (enumeration: c = chapter, s = section, i = index in section) and is followed by the statistical parameters. The latter usually are the number of training set compounds, n, the standard error, s, the correlation coefficient, r, and the F ratio. However, some authors use different notations or even different statistical parameters in their model descriptions. Those parameters are stated but not explained in this book. The original source should be consulted if detailed information is needed.

The discussion above indicates that QPPR models must be selected carefully, considering the structure of query compound and its relationship to the structures represented in the training set. It is often useful to employ different models and to compare the results.

1.5   QUANTITATIVE STRUCTURE – PROPERTY RELATIONSHIPS

A quantitative structure–property relationship (QSPR) is a correlation between a property Y and one or several (m) computable molecular descriptors, $X_1, X_2, \ldots, X_m$:

$$Y = f(X_1, X_2, \ldots, X_m) \qquad (1.5.1)$$

In contrast to a chemical property which can be measured, a molecular descriptor is computed from the molecular structure. Contained in the structural information are the atoms making up the molecule and their spatial arrangement. From the coordinates of the atoms, the geometric attributes (i.e., the size and shape of the molecule) can be deduced. A straightforward example is the molecular mass, which is computed by adding up the masses of the individual atoms making up the molecule and indicated in the elemental composition. The result is accurate since the atomic masses are independent of the chemical bonds with which they are involved. However, the molecular mass reflects few of the geometrical and chemical attributes of a compound and M is therefore a poor predictor for most properties.

Better starting points for developing QSPRs are connection tables that encode the molecular constitution, including information about atom and bond types. Molecular descriptors can be derived from the connection table of the molecule (i.e., the molecular graph) using a set of consistent rules. These descriptors are usually referred to as *molecular connectivity indices* (MCIs). Other notations, such as *graph-theoretical index*, *topological index* or *molecular invariants* are also used. In other cases, ab initio descriptors are employed in QSPRs. For example, a general inter-action property function (GIPF), has been proposed:

$$Y = f(\text{area}, \Pi, \sigma_{\text{tot}}^2, \nu) \qquad (1.5.2)$$

where *area* is the molecular surface area and Π, σ_{tot}^2, and ν are statistically based quantities obtained from the molecular surface electrostatic potentials [27]: Π measures the local polarity, σ_{tot}^2 indicates the variability of the potential on the surface, and ν measures the balance between positive and negative regions. Specific GIPFs have been applied to the estimation of the boiling point, the critical point, and the octanol–water partition coefficient [27].

Computable molecular descriptors differ from experimentally derived properties in two important ways:

1. Their values are inherently precise.
2. They can be derived from the compound structure alone, (i.e., can be determined even for compounds that are not available in pure form).

These characteristics distinguish QPPRs from QSPRs in terms of their statistical evaluation and in terms of their applicability. Note that to estimate the property of interest with a QPPR model, certain other properties of a query compound must be available.

Computable molecular descriptors that occur most frequently in QSPRs in this book are explained in Chapter 2. QSPRs and their statistical parameters are presented in the same way as shown for QPPRs in Section 1.4. Often, QSPR studies apply a set of molecular descriptors to compare their significance for the particular correlation. In this book we present only the most significant QSPRs as judged in the source or by the authors.

1.6 GROUP CONTRIBUTION MODELS

A GCM is a correlation between a property Y and a set of group contributions $G_1, G_2, \ldots, G_m$:

$$Y = f(G_1, G_2, \ldots, G_m) \qquad (1.6.1)$$

where m is the number of group contributions considered in a particular group contribution table. The group contributions are numerical quantities associated with subgraphs (i.e., substructures), of the molecular graph. These subgraphs may be single atoms, atom pairs, or multiatom groups usually associated with functional groups. Group contribution models (GCMs), can accommodate many classes of compounds, whereas QPPR and QSPR models are usually confined to one or only a small number of compound classes.

In this book we use the notation GCM for all three model types: atom contribution, bond contribution, and function-oriented group contribution models. Atom contribution models exhibit a one-to-one correspondence between atoms and contributions (i.e., each atom in a molecule is associated with exactly one contribution). In some schemes separate contributions for H atoms are defined; in other schemes H contributions are included in the contributions of the heavier atoms. An atom group is an atom-centered substructure for which the contribution is specified by the type of the central atom and by its structural environment. Specification of the structural environment depends on the particular atom contribution scheme. Similarly, bond contribution schemes differ with respect to their definition of the bond environment. Function-oriented models use a combination of single- and multiatom groups: for example, halogen atoms (–X) as single-atom groups, a carbonyl group ($>$C=O) as a two-atom group, and a carbamate group [–O–C(=O)–N$<$] as a four-atom group.

In Figure 1.6.1 the conceptual differences between atom, bond, and function-oriented GCMs are demonstrated for propene nitrile. The symbols –, =, and # denote a single, double, and triple bond, respectively. Note, for example, the different representations of the C_{sp} atom in propene nitrile. (See chemistry texts for sp notation.) In the atom contribution approach, C_{sp} is identified as –C#. sp_n indicates

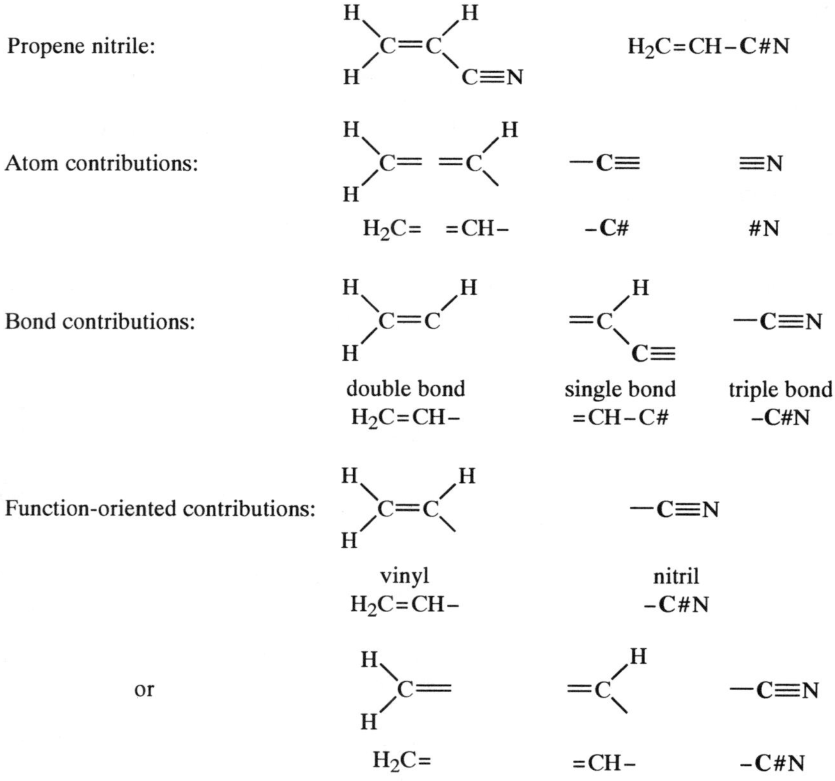

Figure 1.6.1 Various group contribution types for propene nitrile. The SMILES rotation is used to indicate the groups. The C_{sp} hybridize carbon atom is in bold.

the type of hybridization (refer to organic chemistry texts). In the bond contribution approach, the contribution of the C_{sp} carbon is included in the single bond, $=C-C\equiv$, and the triple bond, $-C\equiv N$. In the function-oriented approach, C_{sp} is part of the functional group $-C\equiv N$. The definition of functional groups depends on the particular GCM scheme used. In addition, the subclassification of groups in relation to their structural environment varies among different GCM schemes. Some GCMs distinguish constitutionally equal groups with respect to their occurrence in a chain or a ring. Terminal groups (i.e., substituents), are often differentiated due to their substitution side, such as aromatic ring, nonaromatic ring, or alkyl chain. Use of GCMs always needs thorough consideration of the applied group definitions. There is no GCM standard, and the use of a particular GCM should include a critical observation of the aforementioned characteristics. Generally, application of GCMs to a query molecule requires the following steps:

1. Identification of all groups in the molecule applicable to the particular GCM scheme
2. Calculation of property by employing the function associated with the particular GCM
3. Consideration of certain extra or correction terms that apply to the particular query molecule, such as group interactions (e.g., the intramolecular neutralization of acidic and basic groups)

GCM functions may be linear or nonlinear. Often, GCM schemes include extra terms for diverse structural factors. The differences among GCMs with respect to the form of their group–property function (eq. 1.6.1) are considered in the following.

Linear GCMs   For a molecule consisting of m groups, the linear GCM has the following form:

$$Y = C_0 + \sum G_j \qquad (1.6.2)$$

where C_0 is a model-specific constant, G_j is the contribution for the jth group in the molecule, and the summation is carried over j from 1 to m, m being the number of groups occurring in the molecule. Usually, the GCM equation is expressed as

$$Y = C_0 + \sum n_i G_i \qquad (1.6.3)$$

where G_i is the contribution of the group of type i, n_i is the number of times this group occurs in the molecule, and the summation is carried over all types i. This model assumes the interactions between the groups to be insignificant.

Nonlinear GCMs   For certain properties, model accuracy can be improved by including a quadratic term:

$$Y = C_0 + \sum n_i G_i - \left(\sum n_i G_i \right)^2 \qquad (1.6.4)$$

The method of Lydersen [28] is a GCM of this type to estimate the critical temperature, T_c. Other approaches to non-linear GCMs include the model of Lai et al. [29] for the boiling point, T_b, and the ABC approach [30] to estimate a variety of thermodynamic properties. Further, artificial neural networks have been used to construct nonlinear models for the estimation of the normal boiling point of haloalkanes [31] and the boiling point, critical point, and acentric factor of diverse fluids [32].

Modified GCMs   GCM schemes often consider extra contributions. These contributions can be physicochemical properties, molecular descriptors, and various correction factors. For example, to estimate the critical temperature, T_c, the input of the normal boiling point, T_b, is required by certain GCMs [33]. In this case a typical approach is based on the modification of eq. 1.6.3 as follows:

$$T_c = T_b \left(C_0 + \sum n_i G_i \right)^{-1} \qquad (1.6.5)$$

Computable molecular descriptors have also been introduced in GCMs. The model then becomes

$$Y = C_0 + \sum n_i G_i + X_k \qquad (1.6.6)$$

where X_k is a molecular descriptor of type k. Molecular descriptors of various types are defined in Chapter 2. For example, Suzuki et al. [34] developed a model of the form of eq. 1.6.6 to estimate the air–water partition coefficient, K_{aw}, in which X_k is the molecular connectivity index $^1\chi$. Wang et al. [35] have combined the approach of group contributions with local graph indices for the estimation of T_b.

GCMs are particularly useful in property estimation when used in combination with a factual database and molecular-similarity-based devices. This approach is discussed in the next section.

1.7   SIMILARITY-BASED AND GROUP INTERCHANGE MODELS

Similarity-based and group interchange models (GIMs) use one or several similar compounds (parents) as the starting point for property estimation. The query and the parents may share some structural features and differ in others. It is possible to construct the query from a parent by replacing a portion of the parent with one or several fragments contained in the query. Replacement is a two-step process: a deletion and an insertion. The property estimate is then obtained, knowing the constants of the fragment that is deleted and the fragment that is inserted. A simple example is the construction of chlorobenzene from toluene. This transformation involves the deletion of a methyl and the insertion of a chlorine substituent. Multiple insertions and deletions may be needed. The process of constructing a query is termed *group interchange* (GI) [36]. The GIM accounts for the effect of one or several GI on the property in question. A more involved example is the transformation of 6-bromo-7-phenyl-1-heptenene (**I**) into 4-bromo-5-phenylpentyl oxiran (**II**) (Figure 1.7.1).

Figure 1.7.1 Interchange of the ethenyl and oxiryl groups. The remainder of the molecule, the β-bromo-1-phenyl-*n*-pentyl group, remains intact.

This transformation may be described as an interchange of the ethenyl and the oxiryl group while the remainder of the molecule, the β-bromo-1-phenyl-*n*-pentyl group, remains intact.

The GI approach relates the property, Y_Q, of a query compound to the known property, Y_D, of a database compound, by the equation [36]

$$Y = Y_D + \Delta Y[\text{LNGI}] \qquad (1.7.1)$$

where $\Delta Y[\text{LNGI}]$ is the property difference associated with the structural difference between query and candidate structure, LNGI being the linear notation specifying a particular group interchange. Here, the SMILES notation is used to represent the groups. RE indicates replacement of the first group by the second group between two bars. For example, RE:−C|C=C,C1OC1| is a valid LNGI for the difference between **I** and **II**, representing interchange of the ethenyl group, C=C, by the oxiryl group, C1OC1, as substituents of a methylene group, represented by −C. A much more specific LNGI would be RE:−C(Br)CCC|C=C, C1OC1|, representing the same group interchange but with deeper specification of the substitution side.

Application of the GIM approach in property estimation requires the following steps:

1. Identification of appropriate database structures and their properties
2. Recognition of the structural difference between the query and a selected candidate structure (i.e., assignment of LNGI)
3. Evaluation of the corresponding property difference [i.e., assignment of ΔY(LNGI)].

Steps 1, 2, and 3 are demonstrated below. The rules needed in step 2 to construct unambiguous LNGIs for automatic recognition and evaluation of structural differences are described elsewhere [36]. Step 3 is identical to the derivation of a particular GIM model; examples are given below.

The GIM can be used for a query, Q, if a database is available containing compounds that are structurally similar to the query compounds. Similarity is defined here as sharing a common subgraph. The largest possible subgraph shared by two molecular graphs is denoted as the *maximum common subgraph* (MCS). In Figure 1.7.2 *n*-alkanes are compared with their analogous 2-oxa-alkanes. Similarly, in Figure 1.7.3, *n*-alkanes and 2-thiaalkanes are compared. For each pair the MCS consists of the methyl and the *n*-alkyl group, whereas the interchanged groups are a methylene group and a chalcogen atom. All three pairs in Figure 1.7.2 exhibit the same structural difference: RE:−C|C,O|C. The structural difference for the compounds in Figure 1.7.3 is RE:−C|C,S|C. Based on the

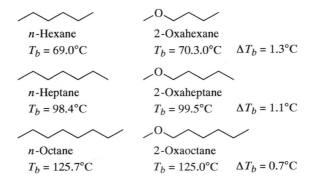

n-Hexane
$T_b = 69.0°C$

2-Oxahexane
$T_b = 70.3.0°C$ $\Delta T_b = 1.3°C$

n-Heptane
$T_b = 98.4°C$

2-Oxaheptane
$T_b = 99.5°C$ $\Delta T_b = 1.1°C$

n-Octane
$T_b = 125.7°C$

2-Oxaoctane
$T_b = 125.0°C$ $\Delta T_b = 0.7°C$

Figure 1.7.2   Evaluation of T_b differences between n-alkanes and 2-oxaalkanes using the data of Balaban [31].

experimental data [31], a mean value for the boiling point differences can be calculated:

$$\Delta T_b[\text{RE: } -C|C, O|C] = 0.57\,°C \tag{1.7.2a}$$

$$\Delta T_b[\text{RE: } -C|C, S|C] = 48.70\,°C \tag{1.7.2b}$$

Equations 1.7.2a and 1.7.2b are each considered a GIM. For example, GIM 1.7.2b can be applied to the problem illustrated in Figure 1.7.3.

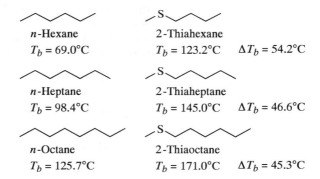

n-Hexane
$T_b = 69.0°C$

2-Thiahexane
$T_b = 123.2°C$ $\Delta T_b = 54.2°C$

n-Heptane
$T_b = 98.4°C$

2-Thiaheptane
$T_b = 145.0°C$ $\Delta T_b = 46.6°C$

n-Octane
$T_b = 125.7°C$

2-Thiaoctane
$T_b = 171.0°C$ $\Delta T_b = 45.3°C$

Figure 1.7.3   Evaluation of T_b differences between n-alkanes and 2-thiaalkanes using the data of Balaban [31].

The usefulness of the approach is illustrated in the example shown in Figure 1.7.4. T_b is known for compound A but not for query X. The two compounds are similar in that the interchange of a methylene group with a thia- ($-S-$) group converts compound A into X. Following eq. 1.7.1, the unknown boiling point is calculated as follows:

$$T_{b,X} = T_{b,A} + 48.70$$

It is important to note that in this estimation procedure the remainder R in A and X can be ignored. R may be complex and application of GCMs to estimate $T_{b,X}$ may require groups and factors encountered in R for which contributions have not yet been incorporated into the particular GCM. Application of the appropriate GIM

Compound A with known $T_{b,A}$ Compound X with $T_{b,X}$ = ?

Figure 1.7.4 Database compound A and query compound X related through RE:$-$C|C,S|C.

circumvents these problems. Note that the validity of a linear GCM implies the validity of corresponding GIMs. In the opposite direction, if a GIM in the form of eq. 1.7.1 cannot be validated for a given LNGI, one can conclude that simple group additivity (eq. 1.6.2) with respect to the interchanged groups does not apply. In this sense, GIM analysis helps to verify the group contribution approach.

GIMs such as 1.7.2a and 1.7.2b represent a constant ΔT_b value consistent with the group additivity assumption underlying the GCM approach. However, T_b, along with many other properties, is not generally constitutive-additive. For such cases, the query–database relationships can be presented in the following general form:

$$Y_Q = f(Y_D, \Delta Y[\text{LNGI}]) \tag{1.7.3}$$

with eq. 1.7.1 as a very special case. Equation 1.7.3 describes a set of functions $Y_Q = f(Y_D)$ where each applies for a particular LNGI. One such function in which Y_Q and Y_C are linearly related will be presented for n-alkanol and n-alkyl β-ethoxypropionate, shown in Figure 1.7.5. Dixon et al. [37] derived the following relationship to estimate T_b of n-alkyl β-ethoxypropionates from T_b of the corresponding n-alkanol:

$$T_{b,\text{ester}} = 105 + 0.85 T_{b,\text{alc}} \qquad \text{with} \quad \text{LNGI} = \text{RE}:-\text{C}|\text{O}, \text{OC}(=\text{O})\text{CCOCC}|$$

$$\tag{1.7.4}$$

where $I_{b,\text{ester}}$, and $T_{b,\text{alc}}$ are the normal boiling points in °C of the ester and the alcohol, respectively.

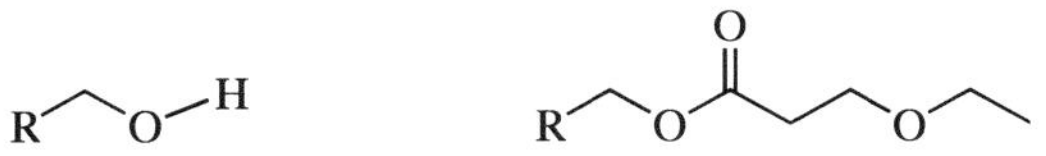

Alcohol with known $T_{b,\text{alc}}$.   Ester with $T_{b,\text{ester}}$ = ?

Figure 1.7.5 n-Alkanol and n-alkyl β-ethoxypropionate as query–database pair.

A final example considers log K_{ow} values for substituted s-triazines as shown in Figure 1.7.6. The structural difference for each of the three pairs is RE:c|O,S|C. The corresponding differences, $\Delta \log K_{\text{ow}}$, range between 0.33 and 0.38 log K_{ow} units, with the following average value:

$$\Delta \log K_{\text{ow}}[\text{RE}:\text{c}|\text{O}, \text{S}|\text{C}] = 0.353$$

If we assume that the K_{ow} value for prometryn is unknown, its value could be estimated from the known K_{ow} values of prometon and either of the pairs atraton–ametryn or terbumeton–terbutryn as follows:

From atraton–ametryn: $\log K_{\text{ow}} = 2.99 + (3.07 - 2.69) = 3.37$
From terbumeton–terbutryn: $\log K_{\text{ow}} = 2.99 + (3.43 - 3.10) = 3.32$

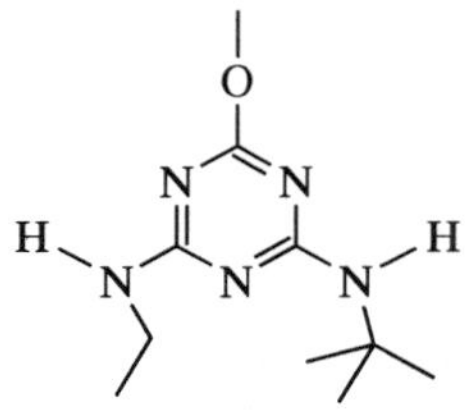

Terbumeton
log K_{ow} = 3.10

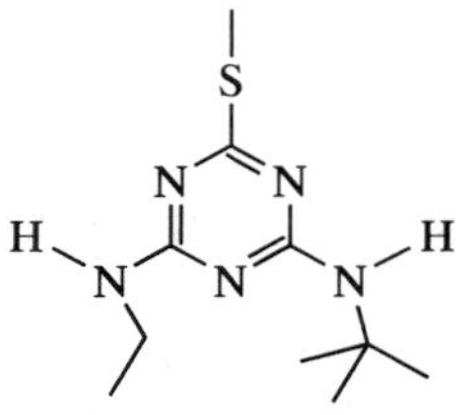

Terbutryn
log K_{ow} = 3.43 Δ log K_{ow} = 0.33

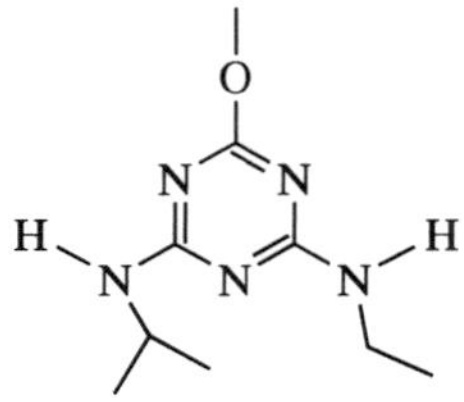

Atraton
log K_{ow} = 2.69

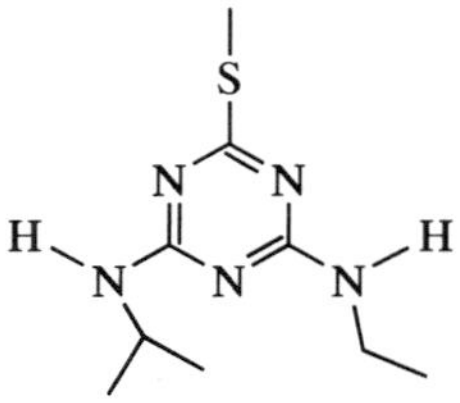

Ametryn
log K_{ow} = 3.07 Δ log K_{ow} = 0.38

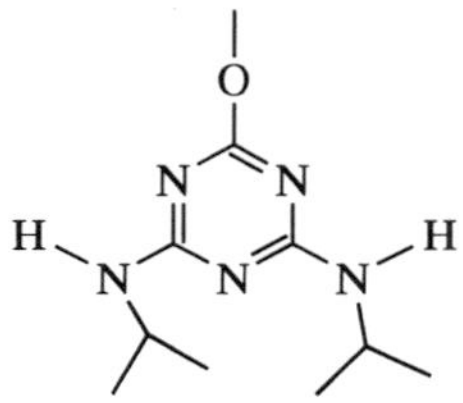

Prometon
log K_{ow} = 2.99

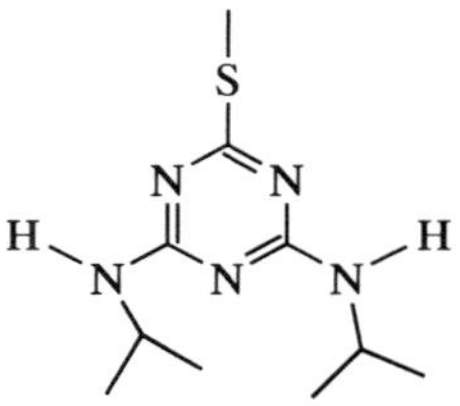

Prometryn
log K_{ow} = 3.34 Δ log K_{ow} = 0.35

Figure 1.7.6 Log K_{ow} value of *s*-triazines [38] and interchange between the methoxy and thiomethoxy groups. (*Source*: Ref. 38.)

where the differences between estimated and observed values are 0.03 and 0.02 log K_{ow} units, respectively. In addition, another group interchange, the replacement of an ethyl by an isopropyl group could be used to estimate K_{ow} of prometryn from ametryn and atraton–prometon:

$$\text{From atraton–prometon:} \qquad \log K_{ow} = 3.07 + (2.99 - 2.69) = 3.37$$

where the difference between estimated and observed values is 0.03 log K_{ow} unit.

Alternative to the use of GIMs, upgraded GCMs have been considered. In the case of *s*-triazines, the group contributions shown in Figure 1.7.7 have been evaluated to enhance the GCM of Hansch and Leo [38]. With some of the available GCMs, however, integration of such large multiatomic groups would conflict with the inherent model logic and definition of group types. In general, however, integration of contributions of large multiheteroatomic groups such as ring systems is highly recommended to improve existing GCMs, although this might seem somewhat

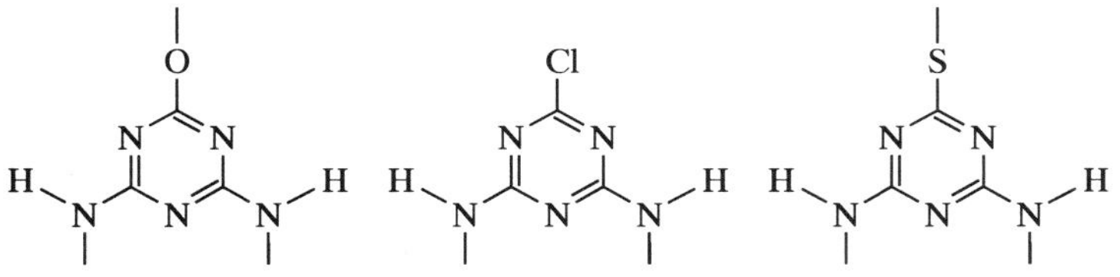

Figure 1.7.7 Group contributions for substituted *s*-triazine substructures. (*Source*: Reprinted from Ref. [38]. Copyright (1991) with permission from Elsevier Science.)

counterproductive to the originally intended simplicity and ease of use in GCM applications.

Application of the GIM approach has the advantage over the GCM approach that the accuracy of the estimated value relies solely on the experimental accuracy of the database compounds and the uncertainty of the contributions for interchanged groups rather than on the uncertainties of all groups that would be needed to estimate a property with a GCM. Especially when a query compound becomes structurally more complex, the number of required group contributions and correction factors increases rapidly and values for a particular correction factor might not yet be available in the GCM scheme considered. In such cases, reliable properties values cannot readily be derived with GCMs unless they are upgraded as aforementioned. The GIM approach reduces significantly the complexity imposed by the variety of factors. Only the factors applying to the interchanged groups have to be accounted for, whereas all those factors confined to the MCS do not have to be considered.

1.8 NEAREST-NEIGHBOR MODELS

The nearest-neighbor (NN) approach relates the property of a query compound, Y_Q, to the properties of k nearest-neighbor (kNN) compounds selected from a database. The general model is

$$Y_Q = f(Y_{D,1}, Y_{D,2}, Y_{D,3}, \ldots, Y_{D,k}) \qquad (1.8.1)$$

where $Y_{D,1}, Y_{D,2}, Y_{D,3}, \ldots, Y_{D,k}$ are the properties of the kNN compounds. Selection of the kNN compounds is based on the measurement of molecular similarity between the query and all database compounds and is usually done using computer algorithms. Molecular similarity can be measured in various ways. For example, Basak and Grunwald [39] compare the performance of two different similarity measurements: atom pair (AP) similarity and topological index (TI) similarity. The AP similarity is based on APs that are derived from the distance matrix of the molecular graph (Section 2.2). The TI similarity is derived from a set of computable molecular descriptors such as those discussed in Chapter 2. In-depth discussion of molecular similarity measurements can be found elsewhere (e.g., [36,40–46]).

The kNN compounds used to estimate the query property are those k database compounds that exhibit the greatest similarity to the query compound. Basak and Grunwald [42] and Basak et al. [47] use the mean of the kNN property values

$Y_{D,1}, Y_{D,2}, Y_{D,3}, \ldots, Y_{D,k}$ as an estimate of Y_Q considering $Y = \log K_{ow}$ and $Y = T_b$, respectively.

$$Y_Q = \frac{1}{k}(Y_{D,1} + Y_{D,2} + Y_{D,3} + \cdots + Y_{D,k}) \qquad (1.8.2)$$

The *k*NN approach has also been applied to compound classification. For example, Drefahl [48] has discussed the *k*NN approach to discriminate chlorinated organics with respect to their volatility from aqueous solutions. Varmuza [49] has described *k*NN classification as a standard method in pattern recognition and provides references to its use in spectra interpretation.

Performance of the *k*NN approach depends on (1) the content of the database, and (2) the particular method employed for similarity measurement. Measurement of the molecular similarity quantifies the similarity between two compounds and thus allows a ranking of the database compounds with respect to any query compound. This quantification and ranking process is based solely on molecular structure information on the query and database compounds. The mean value approach is justified if one assumes that the *k*NN compounds are the most similar compounds for a query with respect to both structure and property. Often, however, compounds evaluated as being structurally similar exhibit large differences in their physico-chemical properties (see, e.g., Figure 10.4.1). Therefore, including GIMs (Section 1.7) will significantly improve property estimation with the *k*NN approach. Drefahl and Reinhard [36] have demonstrated the combined use of the *k*NN and GIM approach for the estimation of $\log K_{ow}$. Further systematic investigation of the *k*NN / GIM approach is key in future structure–property studies and the computerized development of accurate, integrated, and informative estimation modules.

1.9 METHODS TO ESTIMATE TEMPERATURE-DEPENDENT PROPERTIES

Most QPPRs, QSPRs, or GCMs yield properties at a reference temperature. Temperature-dependent property estimation is often restricted to compounds for which the compound-specific temperature coefficients in thermodynamic or empirical relationships have been evaluated. Compound-specific temperature functions have the following general form:

$$Y = f(T, a_0, a_1, \ldots, a_m) \qquad (1.9.1)$$

where $a_0, a_1, \ldots, a_m$ are the temperature coefficients. Examples of density, viscosity, and air–water partition coefficients are given in Appendixes B through D.

Many thermodynamic models can be considered as property–temperature–property relationships:

$$Y = f(T, P_1, P_2, \ldots, P_{1m}) \qquad (1.9.2)$$

which are equivalent to eq. 1.4.1 except for T as one additional independent variable. The standard source of information on such methods is *The Properties of Gases and Liquids* [50].

In certain cases structure–temperature–property relationships have been developed that allow the estimation of a property as a function of both structure and temperature but do not require any additional compound properties. The general model is

$$Y = f(T, \text{molecular structure}) \qquad (1.9.3)$$

GCMs that allow temperature-dependent property estimations are specific cases of model 1.9.3:

$$Y = f(T, G_1, G_2, \ldots, G_m) \qquad (1.9.3a)$$

where $G_1, G_2, \ldots, G_m$ are group contributions as encountered in eq. 1.6.1.

Selected compound-specific functions, property–temperature–property relationships, or structure–temperature–property relationships are supplied and discussed in this book for density (Section 3.5), refractive index (Section 4.5), surface tension (Section 5.4), viscosity (Section 6.4), vapor pressure (Section 7.4), enthalpy of vaporization (Section 8.5), aqueous solubility (Section 11.8), and air–water partition coefficients (Section 12.5).

REFERENCES

1. Auer, C. J. M., J. V. Nabholz, and K. P. Baetke, *Mode of Action and the Assessment of Chemical Hazards in the Presence of Limited Data: Use of Structure–Activity Priorities*, 1990. Washington, DC: National Academy Press.

2. Weininger, D., SMILES, A Chemical Language and Information System: 1. Introduction to Methodology and Encoding Rules. *J. Chem. Inf. Comput. Sci.*, 1988: **28**, 31–36.

3. Weininger, D., A. Weininger, and J. L. Weininger, SMILES: 2. Algorithm for Generation of Unique SMILES Notation. *J. Chem. Inf. Comput. Sci.*, 1989: **29**, 97–101.

4. Weininger, D., SMILES: 3. DEPICT: Graphical Depiction of Chemical Structures. *J. Chem. Inf. Comput. Sci.*, 1990: **30**, 237–243.

5. Amidon, G. L. and S. T. Anik, Application of the Surface Area Approach to the Correlation and Estimation of Aqueous Solubility and Vapor Pressure: Alkyl Aromatic Hydrocarbons. *J. Chem. Eng. Data*, 1981: **26**, 28–33.

6. van Haelst, A. G., et al., Determination of *n*-Octanol/Water Partition Coefficients of Tetrachlorobenzyltoluenes Individually and in a Mixture by the Slow Stirring Method. *Chemosphere*, 1994: **29**, 1651–1660.

7. Testa, B., *Principles of Organic Stereochemistry*. 1979. New York: Marcel Dekker.

8. Maas, B., A. Dietrich, and A. Mosandl, Enantioselective Capillary Gas Chromatography— Olfactometry in Essential Oil Analysis. *Naturwissenschaften*, 1993: **80**, 470–472.

9. Pfaffenberger, B., et al., Gas Chromatographic Separation of the Enantiomers of Bromocyclene in Fish Samples. *Chemosphere*, 1994: **29**, 1385–1391.

10. Schurig, V., and A. Glausch, Enantiomer Separation of Atropisomeric Polychlorinated Biphenyls (PCBs) by Gas Chromatography on Chirasil-Dex. *Naturwissenschaften*, 1993: **80**, 468–469.

11. Eardman, D., Chiral Engineering Breaks Through the Looking Glass. *Chem. Eng.*, 1993: Oct., 35–39.

12. Grummitt, O., and D. Marsh, Geometric Isomers of 1-Phenyl-1,3-butadiene. *J. Am. Chem. Soc.*, 1949: **71**, 4157.

13. Randić, M., in *Design of Molecules with Desired Properties: A Molecular Similarity Approach to Property Optimization Concepts and Applications of Molecular Similarity*. M. A. Johnson and G. M. Maggiora., Editors, 1990. New York: Wiley.

14. Trinajstić, N., Enumeration of Isomeric Acyclic Structures, in *Computational Chemical Graph Theory: Characterization, Enumeration and Generation of Chemical Structures by Computer Methods*, S. Nikolic' et al., Editors, 1991. Ellis Horwood: New York.

15. Coffman, D. D., The Number of Structurally Isomeric Hydrocarbons of the Acetylene Series. *J. Am. Chem. Soc.*, 1933: **55**, 252–253.

16. Coffman, D. D., The Number of Stereoisomeric and Non-stereoisomeric Alkenes. *J. Am. Chem. Soc.*, 1933: **55**, 695–698.

17. Henze, H. R., and C. M. Blair, The Number of Structurally Isomeric Hydrocarbons of the Ethylene Series. *J. Am. Chem. Soc.*, 1933: **55**, 680–686.

18. Perrin, D. M. S. J., and J. R. S. J. VandeVelde, Of Men and Marigolds: Counting the Quaterthienyls. *J. Chem. Educ.*, 1992: **69**, 719–723.

19. Hall, W. P. and E. E. Reid, A Series of α,ω-Dimercaptans. *J. Am. Chem. Soc.*, 1943: **65**, 1466–1468.

20. Wallace, J. M., Jr. and J. E. Copenhaver, Anhydrides of the Normal Aliphatic Saturated Monobasic Acids. *J. Am. Chem. Soc.*, 1941: **63**, 699–700.

21. Hoback, J. H., D. O. Parsons, and J. F. Bartlett, Some Esters of Normal Aliphatic Alcohols and Acids. *J. Am. Chem. Soc.*, 1943: **65**, 1606–1607.

22. Armstrong, M. D., and J. E. Copenhaver, Alkyl *p*-Nitrobenzoates. *J. Am. Chem. Soc.*, 1943: **65**, 2252–2253.

23. Reid, E. E., and E. Wilson, Some Mono- and Di-alkyl Ethers of Stilboestrol. *J. Am. Chem. Soc.*, 1942: **64**, 1625–1626.

24. Burrows, H. D., Studying Odd–Even Effects and Solubility Behavior Using $\alpha,\omega=$ Dicarboxylic Acids. *J. Chem. Educ.*, 1992: **69**, 69–73.

25. Campbell, C., Simple Linear Regression, in *Handbook of Chemical Property Estimation*, W. J. Lyman, W. F. Reehl, and D. H. Rosenblatt, Editors, 1990. Washington, DC: American Chemical Society.

26. Anonymous, Recommendations for Reporting the Results of Correlation Analysis in Chemistry using Regression Analysis. *Quant. Struct.-Act. Relat.*, 1985: **4**, 29.

27. Murray, J. S., et al., Statistically-Based Interaction Indices Derived from Molecular Surface Electrostatic Potentials: A General Interaction Properties Function (GIPF). *J. Mol. Struct. (Theochem.)*, 1994: **307**, 55–64.

28. Rechsteiner, C. E., Boiling Point, in *Handbook of Chemical Property Estimation*, W. J. Lyman, W. F. Reehl, and D. H. Rosenblatt, Editors, 1990. Washington, DC: American Chemical Society.

29. Lai, W. Y., D. H. Chen, and R. N. Maddox, Application of a Nonlinear Group-Contribution Model to the Prediction of Physical Constants: 1. Predicting Normal Boiling Points with Molecular Structure. *Ind. Eng. Chem. Res.*, 1987: **26**, 1072–1079.

30. Constantinou, L., S. E. Prickett, and L. Mavrovouniotis, Estimation of Thermodynamic and Physical Properties of Acyclic Hydrocarbons Using the ABC Approach and Conjugation Operators. *Ind. Eng. Chem. Res.*, 1993: **32**, 1734–1746.

31. Balaban, A. T., et al., Correlation Between Structure and Normal Boiling Points of Haloalkanes C1–C4 Using Neural Networks. *J. Chem. Inf. Comput. Sci.*, 1994: **34**, 1118–1121.

32. Lee, M. J., and J. -T. Chen, Fluid Property Predictions with the Aid of Neural Networks. *Ind. Eng. Chem. Res.*, 1993: **32**, 995–997.

33. Klincewicz, K. M., and R. C. Reid, Estimation of Critical Properties with Group Contribution Methods. *AIChE*, 1984: **30**, 137–142.

34. Suzuki, T., K. Ohtaguchi, and K. Koide, Application of Principal Components Analysis to Calculate Henry's Constant from Molecular Structure. *Comput. Chem.*, 1992: **16**, 41–52.

35. Wang, S., G. W. A. Milne, and G. Klopman, Graph Theory and Group Contributions in the Estimation of Boiling Points. *J. Chem. Inf. Comput. Sci.*, 1994: **34**, 1242–1250.

36. Drefahl, A., and M. Reinhard, Similarity-Based Search and Evaluation of Environmentally Relevant Properties for Organic Compounds in Combination with the Group Contribution Approach. *J. Chem. Inf. Comput. Sci.*, 1993: **33**, 886–895.

37. Dixon, M. B., C. E. Rehberg, and C. H. Fisher, Preparation and Physical Properties of n-Alkyl-β-Ethoxypropionates. *J. Am. Chem. Soc.*, 1948: **70**, 3733–3738.

38. Finizio, A., et al., Different Approaches for the Evaluation of K_{ow} for s-Triazine Herbicides. *Chemosphere*, 1991: **23**, 801–812.

39. Basak, S. C., and G. D. Grunwald, Use of Topological Space and Property Space in Selecting Structural Analogs, in *Mathematical Modelling and Scientific Computing*, in press. 1998.

40. Basak, S. C., G. J. Niemi, and G. D. Veith, Optimal Characterization of Structure for Prediction of Properties. *J. Math. Chem.*, 1990: **4**, 185–205.

41. Basak, S. C., G. J. Niemi, and G. D. Veith, Predicting Properties of Molecules Using Graph Invariants. *J. Math. Chem.*, 1991: **7**, 243–272.

42. Basak, S. C. and G. D. Grunwald, Estimation of Lipophilicity from Molecular Structural Similarity. *New J. Chem.*, 1995: **19**, 231–237.

43. Johnson, M., S. Basak, and G. Maggiora, A Characterization of Molecular Similarity Methods for Property Prediction. *Math. Comput. Model.*, 1988: **11**, 630–634.

44. Johnson, M. A., A Review and Examination of the Mathematical Spaces Underlying Molecular Similarity Analysis. *J. Math. Chem.*, 1989: **3**, 117–145.

45. Johnson, M. A., and G. M. Maggiora, *Concepts and Applications of Molecular Similarity*, 1990. New York: Wiley.

46. Willett, P., V. Winterman, and D. Bawden, Implementation of Nearest-Neighbor Searching in an Online Chemical Structure Search System. *J. Chem. Inf. Comput. Sci.*, 1986: **26**, 36–41.

47. Basak, S. C., B. D. Gute, and G. D. Grunwald, Estimation of Normal Boiling Points of Haloalkanes Using Molecular Similarity. *Croat. Chim. Acta*, 1966: **69**, 1159–1173.

48. Drefahl, A., Modellentwicklungen zur Vorhersage des Umweltverhaltens organischer Verbindungen auf der Basis computergestützter Struktur / Eigenschafts-Transformationen, 1988. Ph.D. dissertation, Technical University of Munich, Department of Organic Chemistry.

49. Varmuza, K., *Pattern Recognition in Chemistry*, 1980. Lecture Notes in Chemistry 21. Berlin: Springer-Verlag.

50. Reid, R. C., J. M. Prausnitz, and B. E. Poling, *The Properties of Gases and Liquids*, 4th ed., 1988, New York: McGraw-Hill.

CHAPTER 2

COMPUTABLE MOLECULAR DESCRIPTORS

2.1 INTRODUCTION

Computable molecular descriptors, X, are invariants (constants) that are calculated from the topological information contained in the structure or graph of a molecule [1–3]. Topological information of a molecule comprises the position and sometimes the type of the atoms defined in relation to the bonds that connect them. Such topological descriptors correlate with certain compound properties and activities. Figure 2.1.1 shows two examples for molecular descriptor development: First, the adjacency matrix representing the basic graph is constructed. In the simplest case, the graph does not show hydrogens and specify the type of atoms present. Then, an algorithm is applied to the matrix, which assigns a numerical value or vector, X to the matrix. The values X for a population of compounds can then be correlated with property data. Alternatively, matrices can be constructed that represent the chemical information of the structure, such as the identity of the atoms present. The latter yields "chemically informed" molecular descriptors.

In graph theory, graphs are defined by an ordered pair consisting of two sets, V and R:

$$G = [V, R] \qquad (2.1.1)$$

The elements of V are called *vertices* and the elements of R are called *edges*. In the context of chemical graph theory, vertices and edges represent atoms and bonds, respectively. In basic graphs, the atom types and bond orders are not specified. Graph-theoretical invariants reflect structural characteristics, such as connectivity, connectedness, or branching. A number of graph-theoretical invariants have been defined, but only a small selection are discussed here. Applications of molecular descriptors for predicting the biological and environmental behavior of organic compounds are discussed by Brezonik [4].

Three different variables are used for calculating molecular descriptors: indicator variable, count variable, and graph-theoretical indices, as described below. Molecular

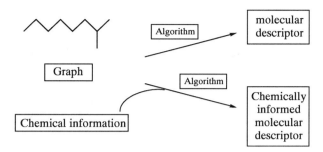

Figure 2.1.1 Development of molecular descriptors.

descriptors can therefore be computed on the basis of structure information alone without having to rely on measured values.

Indicator Variables *Indicator variables* are binary variables having a value of either 0 or 1, "indicating" if a particular structural or substructural characteristic is present or is missing in the molecular graph. Indicator variables are used by some authors in QSPR equations presenting correlations between a property and descriptors, including indicator variables. However, indicator variables are redundant. Instead of using an indicator, a QSPR model can be split into two separate models, one for all those compounds for which the indicator variable is zero, and one for all those compounds for which the indicator variable is 1.

Count Variables *Count Variables* have an integer value, "counting" the number of times a particular structural feature is present in the molecule. Group contribution models (GCMs), are based on count variables (counts of group occurrences) along with the associated contribution value. GCMs are discussed in Section 1.6. Figure 2.1.2 shows selected count variables illustrated with 2-chloro-1,3-butadiene.

Graph-Theoretical Indices *Graph-theoretical indices* are invariants of the molecular graph (Figure 2.1.3). The key to calculate graph-theoretical indices is the adjacency matrix **A**. **A** is derived from the hydrogen-suppressed molecular graph G of a compound. For example *N,N*-dimethyl acetamide (Ia), tetrachloroethene (Ib), and 2,3-dimethylbutane (Ic) have the same hydrogen-suppressed molecular graph (G-I). A molecular graph representation such as G-I is called a *hydrogen-suppressed graph*

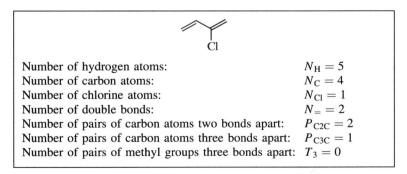

Number of hydrogen atoms:	$N_H = 5$
Number of carbon atoms:	$N_C = 4$
Number of chlorine atoms:	$N_{Cl} = 1$
Number of double bonds:	$N_= = 2$
Number of pairs of carbon atoms two bonds apart:	$P_{C2C} = 2$
Number of pairs of carbon atoms three bonds apart:	$P_{C3C} = 1$
Number of pairs of methyl groups three bonds apart:	$T_3 = 0$

Figure 2.1.2 Count variables for 2-chloro-1,3-butadiene.

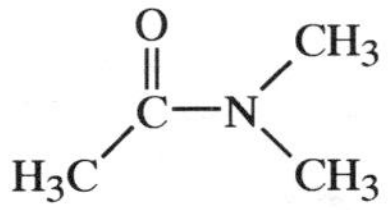

Figure 2.1.3 Examples of a molecular graph.

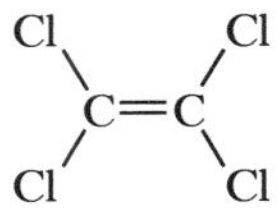

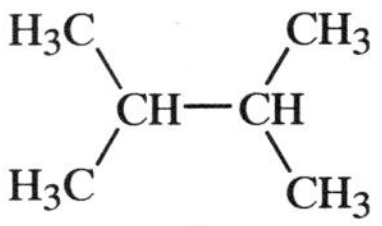

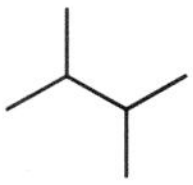

because only nonhydrogen atoms are presented. Atom and bond type information is also suppressed, reducing the molecule to a set of vertices and edges. By labeling the vertices of G-I from 1 to 6, **A** of G-I is given as follows:

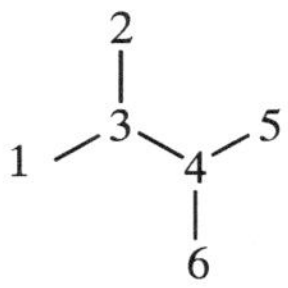

G-I with labeled vertices

Generally, **A** is defined as [5]

$$(\mathbf{A})_{ij} = \begin{cases} 1 & \text{if vertices } i \text{ and } j \text{ are adjacent} \\ 0 & \text{if either } i = j \text{ or } i \text{ and } j \text{ are nonadjacent} \end{cases} \qquad (2.1.2)$$

A is a square $n \times n$ symmetric matrix, where n is the number of nonhydrogen atoms in the molecule. The adjacency matrix of G-I is as follows:

	1	2	3	4	5	6
1	0	0	1	0	0	0
2	0	0	1	0	0	0
3	1	1	0	1	0	0
4	0	0	1	0	1	1
5	0	0	0	1	0	0
6	0	0	0	1	0	0

$$\mathbf{A}(\text{G-I})$$

Cyclomatic Number of G

The *cyclomatic number* μ is a molecular descriptor that indicates the number of bonds that must be broken to obtain an aliphatic structure. Defined in terms of graph theory, μ of graph G is equal to the minimum number of edges that must be removed from G to transform it into G of a related acyclic graph [6,7]. Removing an appropriate edge is the graph-theoretic equivalent to opening a ring by breaking a bond. For acyclic molecules (represented as *trees*) μ is equal to zero; for monocyclic compounds, μ is equal to 1.

2.2   MATRICES DERIVED FROM THE ADJACENCY MATRIX

A is used to generate the distance matrix **D** [8–10]. **D** is a square $n \times n$ symmetric matrix in which the entry $(\mathbf{D})_{ij}$ indicates the distance between vertices i and j, where the distance is the minimum number of edges between i and j. The maximum distance in **D** is denoted as d_{max}. **D** of G-I is

$$
\begin{array}{c|cccccc}
 & 1 & 2 & 3 & 4 & 5 & 6 \\
\hline
1 & 0 & 2 & 1 & 2 & 3 & 3 \\
2 & 2 & 0 & 1 & 2 & 3 & 3 \\
3 & 1 & 1 & 0 & 1 & 2 & 2 \\
4 & 2 & 2 & 1 & 0 & 1 & 1 \\
5 & 3 & 3 & 2 & 1 & 0 & 2 \\
6 & 3 & 3 & 2 & 1 & 2 & 0 \\
\end{array}
$$

$$\mathbf{D}(\text{G-I})$$

with $d_{max} = 3$. The entry of 2 for vertex 3 and 6 in **D**(G-I), for example, indicates that the distance between vertices 3 and 6 in G-I is 2 (i.e., that 2 edges lie between vertex 3 and 6).

Further, **A** is used to generate the valence vector, **v** [11]. This vector indicates the vertex degree for each atom in G:

$$(\mathbf{v})_i = \sum (\mathbf{A})_{i,j} \tag{2.2.1}$$

where the summations is over all j. Based on **D**, a distance sum vector, $\mathbf{d_S}$, has been defined [12]:

$$(\mathbf{d_S})_i = \sum (\mathbf{D})_{i,j} \tag{2.2.2}$$

where again the summation is over all j. Using the atom enumeration in G-I, the corresponding valence and distance sum vectors are

$$\mathbf{v} = \begin{bmatrix} 1 & 1 & 3 & 3 & 1 & 1 \end{bmatrix}$$
$$\mathbf{d_S} = \begin{bmatrix} 11 & 11 & 7 & 7 & 11 & 11 \end{bmatrix}$$

The value of 3 for the third and fourth atoms in **v**, for example, indicates that both atoms have three neighbor atoms. A value of 1 indicates terminal atoms.

Estrada [13] proposes the use of the edge-adjacency matrix, **E**, to develop new molecular descriptors. For example, G-I has five edges, which can be labeled in the following way:

G-I with labeled edges

By considering each edge as a vertex, and connecting a pair of these vertices if they present two incident edges in G-I, the following edge-adjacency graph E-I is obtained:

E-I corresponding to G-I

Then the *edge-adjacency matrix* **E** of G-I is derived as the *normal adjacency matrix* of E-I:

	1	2	3	4	5
1	0	1	1	0	0
2	1	0	1	0	0
3	1	1	0	1	1
4	0	0	1	0	1
5	0	0	1	1	0

E(G-I)

In general terms, **E** is defined as follows [13]:

$$(\mathbf{E})_{fg} = \begin{cases} 1 & \text{if edges } f \text{ and } g \text{ are incident} \\ 0 & \text{otherwise} \end{cases} \qquad (2.2.3)$$

E is a square $m \times m$ symmetric matrix, where m is the number of bonds in the molecule. Further, an edge degree vector **e** can be derived as follows [13]:

$$(\mathbf{e})_f = (\mathbf{v})_i + (\mathbf{v})_j - 2 \qquad (2.2.4)$$

where $(\mathbf{e})_f$ is the edge degree of edge f, f being the edge between vertices i and j, and $(\mathbf{v})_i$ and $(\mathbf{v})_j$ are the vertex degrees of vertices i and j, respectively. The edge degree vector of G-I is

$$\mathbf{e} = [2 \quad 2 \quad 4 \quad 2 \quad 2]$$

Diudea and colleagues introduced the branching matrix, **B** [14]. **B** can be derived directly from **v** and **D** as follows:

$$(\mathbf{B})_{ik} = \sum_j [(\mathbf{v})_j \cdot T_{ij}] \qquad (2.2.5)$$

where the summation is carried out over vertices j and T_{ij} is taken as

$$T_{ij} = \begin{cases} 1 & \text{if } (\mathbf{D})_{ij} = k - 1 \\ 0 & \text{if } (\mathbf{D})_{ij} \neq k - 1 \end{cases} \qquad (2.2.5a)$$

The index k runs from 1 to $d_{max} + 1$. Note that $(\mathbf{B})_{i1} = (\mathbf{v})_i$. Generally, $(\mathbf{B})_{it}$ is the sum of the vertex degree for all vertices situated at distance $t - 1$ from the vertex i. **B** of G-I is shown as an example:

$$
\begin{array}{c|cccc}
 & \mathbf{1} & \mathbf{2} & \mathbf{3} & \mathbf{4} \\
\hline
\mathbf{1} & 1 & 3 & 4 & 2 \\
\mathbf{2} & 1 & 3 & 4 & 2 \\
\mathbf{3} & 3 & 5 & 2 & 0 \\
\mathbf{4} & 3 & 5 & 2 & 0 \\
\mathbf{5} & 1 & 3 & 4 & 2 \\
\mathbf{6} & 1 & 3 & 4 & 2 \\
\end{array}
$$

$$\mathbf{B}(\text{G-I})$$

The value $(\mathbf{B})_{1,4} = 2$ in $\mathbf{B}(\text{G-I})$, for example, represents atom 1 in G-I. Atoms 5 and 6 have a distance of $4 - 1 = 3$ from atom 1. Since the vertex degree of 5 and 6 is 1 for both atoms, $(\mathbf{B})_{1,4} = 1 + 1 = 2$.

In analogy to $\mathbf{B}$, Balaban and Diudea [12] introduced the *regressive distance sum matrix*, $\mathbf{R}$, which is derived by substituting $\mathbf{d_S}$ for $\mathbf{v}$:

$$(\mathbf{R})_{ik} = \sum [(\mathbf{d_S})_j \cdot T_{ij}] \tag{2.2.6}$$

where the summation is carried out over vertices j, and T_{ij} is taken as

$$
T_{ij} = \begin{cases} 1 & \text{if } (\mathbf{D})_{ij} = k - 1 \\ 0 & \text{if } (\mathbf{D})_{ij} \neq k - 1 \end{cases} \tag{2.2.6a}
$$

The index k runs from 1 to $d_{\max} + 1$. Note that $(\mathbf{R})_{i1} = (\mathbf{d_S})_i$.

A variety of molecular descriptors are derived directly from matrices $\mathbf{A}$, $\mathbf{D}$, $\mathbf{E}$, $\mathbf{B}$, and $\mathbf{R}$. Some of them are introduced in the next section.

2.3 DESCRIPTORS DERIVED FROM MATRICES A, D, E, B, AND R

Wiener Index   The *Wiener index*, W, is defined as the sum of topological distances in G [15], considering either the upper right or lower left triangle of $\mathbf{D}$:

$$W = \sum_{i<j} (\mathbf{D})_{i,j} \tag{2.3.1}$$

where the summation is taken from $i = 1$ to n for all i less than j. W of G-I is calculated as follows:

$$
\begin{aligned}
W &= (2 + 1 + 2 + 3 + 3) + (1 + 2 + 3 + 3) + (1 + 2 + 2) + (1 + 1) + 2 \\
&= \quad\quad 11 \quad\quad + \quad\quad 9 \quad\quad + \quad\quad 5 \quad + \quad 2 \quad + 2 \\
&= 29
\end{aligned}
$$

Harary Index   The *Harary index*, H, is defined as the sum of the squares of the reciprocal nonzero distances in $\mathbf{D}$ [6,7]:

$$H = 0.5 \sum_i \sum_j (\mathbf{D})_{i,j}^{-2} \tag{2.3.2}$$

where the first summation is taken from $i = 1$ to n and the second summation is taken from $j = 1$ to n.

Molecular Topological Index   The *molecular topological index* (MTI) is derived from $\mathbf{v}'$, calculated as the sum of $\mathbf{D}$ and $\mathbf{A}$ premultiplied by the row vector $\mathbf{v}$ [11]:

$$\mathbf{v}' = \mathbf{v}(\mathbf{D} + \mathbf{A}) \qquad (2.3.3)$$

MTI is the sum of all elements $(\mathbf{v})_i$ in $\mathbf{v}'$:

$$\mathrm{MTI} = \sum (\mathbf{v}')_i \qquad (2.3.4)$$

The MTI of G-I is calculated as follows:

$$\mathrm{MTI} = 20 + 20 + 14 + 14 + 20 + 20 = 108$$

Balaban Index   The *Balaban index*, J, has been introduced as the average-distance sum defined as the sum of the squares of the reciprocal nonzero distances in $\mathbf{D}$ [6,7]:

$$J = \frac{m}{(\mu + 1) \sum [(\mathbf{ds})_i \cdot (\mathbf{ds})_j]^{0.5}} \qquad (2.3.5)$$

where m is the number of edges, μ is the cyclomatic number, and the summation is taken over all edges (i, j).

Edge-Adjacency Index   The *edge-adjacency index*, ε, has been defined by Estrada [13] as follows:

$$\epsilon = \sum [(\mathbf{e})_f \cdot (\mathbf{e})_g]_{np}^{-1/2} \qquad (2.3.6)$$

where the sum is over all np pairs of adjacent edges f and g and $(\mathbf{e})_f$ and $(\mathbf{e})_g$ are the edge degrees of edges f and g, respectively. ε of G-I is computed as follows:

$$\epsilon = 4(2 \cdot 4)^{-1/2} + 2(2 \cdot 2)^{-1/2}$$
$$= 2^{1/2} + 1$$
$$= 2.4142$$

Charge Indices   *Charge indices G_k and J_k* are calculated from the matrix product $\mathbf{M} = (\mathbf{A} \times \mathbf{D}^*)$, where $\mathbf{D}^*$ is the *inverse square distance matrix* [16]:

$$(\mathbf{D}^*)_{ij} = \begin{cases} [(\mathbf{D})_{ij}]^{-2} & \text{if } (\mathbf{D})_{ij} \neq 0 \quad \text{(i.e., if } i \neq j) \\ 0 & \text{if } (\mathbf{D})_{ij} = 0 \quad \text{(i.e., if } i = j) \end{cases} \qquad (2.3.7)$$

Charge terms (CTs) are defined as $CT_{ij} = (\mathbf{M})_{ij} - (\mathbf{M})_{ji}$. Then for each distance k that occurs in $\mathbf{D}$, G_k is given by the following equation:

$$G_k = \sum_{i=1, j=i+1}^{i=N-1, j=N} |CT_{ij}| \delta(k, (\mathbf{D})_{ij}) \tag{2.3.8}$$

where N is the number of nonhydrogen atoms in the molecule and $\delta(k, (\mathbf{D})_{ij})$ equals 1 if $k = (\mathbf{D})_{ij}$ and 0 otherwise. J_k is defined as follows:

$$J_k = \frac{G_k}{N-1} \tag{2.3.9}$$

Correlations between charge indices and properties such as the dipole moment of hydrocarbons, the boiling point alkanes and alkanols, and the enthalpy of vaporization of alkanes have been studied [16].

Information-Theoretical Indices   Information theory has been employed to define topological indices based on the Shannon equation [17]:

$$I_{\text{mean}} = \frac{I_{\text{total}}}{N} = -\sum \frac{N_i}{N} \log_2 \frac{N_i}{N} \tag{2.3.10}$$

where I_{mean} is the mean information content, I_{total} the total information content, N the number of elements in a given set, and N_i the number of elements in the ith subset. I_{mean} is also denoted as the information content (IC). The summation is carried over all sets of elements. Any of the vectors or matrices defined above, for example, can serve as a set. The subsets are then given by grouping the elements of the vector or matrix with the same value, respectively. For example, two subsets with four and two elements are obtained for $\mathbf{v}$ of G-I:

$$\mathbf{v} = \begin{bmatrix} 1 & 1 & 3 & 3 & 1 & 1 \end{bmatrix}$$

Set from $\mathbf{v}$: $\{\{1,1,1,1\}, \{3,3\}\}$

I_{mean} of $\mathbf{v}$: $-\left[\left(\frac{4}{6}\right) \log_2 \left(\frac{4}{6}\right) + \left(\frac{2}{6}\right) \log_2 \left(\frac{2}{6}\right) \right] = 1.252$

There are a vast number of possible recipes to derive information-theoretical indices. In-depth discussions of selected information-theoretical indices, also referred to as information content indices (ICI), appear elsewhere [17–19].

Determinants and Eigenvalues of A and D   Several molecular descriptors are defined from determinants or eigenvalues of $\mathbf{A}$, $\mathbf{D}$, and $\mathbf{A} + \mathbf{D}$. The interested reader is referred elsewhere [20,21].

Indices Based on Atom-Pair Weighting   Atom pairs (i, j) in a molecular graph can be assigned a weight w_{ij}. The topological distance is a possible weighting scheme: $w_{ij} = (\mathbf{D})_{ij}$. A general index can be defined as the sum of the kth power of the atom-pair weights:

$$S_{AP}^{(k)} = \sum_p (w_{ij})_p^k \qquad (2.3.11)$$

where p is the label for an atom pair (i,j) and the summation is carried over all pairs. Many known indexes can be derived as special cases of definition (2.3.11) [22]. For example, the Wiener index is $S_{AP}^{(1)}/2$ where $w_{ij} = (\mathbf{D})_{ij}$. The first-order molecular connectivity index, $^1\chi$, also called Randic index [6,7], equals $S_{AP}^{(-0.5)}$ with $w_{ij} = 0$ for nonadjacent atom pairs and $w_{ij} = \delta_i \delta_j$ for adjacent atom pairs:

$$^1\chi = \sum (\delta_i \delta_j)^{-0.5} \qquad (2.3.12)$$

where δ_i and δ_j are delta values for atoms i and j, respectively [23]. Delta values are discussed in the next section.

2.4   DESCRIPTORS BASED ON ADDITIONAL INFORMATION

Molecular descriptors derived solely from $\mathbf{D}$, $\mathbf{E}$, $\mathbf{B}$, and $\mathbf{R}$ discriminate between different basic graphs. They do not, however, differentiate between molecules such as I, II, and III with the same basic graph but with differences in their types of atoms, bonds, or stereo- and quantum-chemical features. In the remaining part of this section, a few approaches that extend basic graph descriptors to "chemically informed" descriptors are introduced.

Delta Value Schemes and Molecular Connectivity Indices   A *delta* (δ or δ^v) *value* is an atomic descriptor for nonhydrogen atoms in a molecular graph. The superscript-free delta value, δ_i, is defined as the number of adjacent nonhydrogen atoms of atom i:

$$\delta_i = (\mathbf{v})_i \qquad (2.4.1)$$

The valence-delta value, δ^v, is calculated from the atomic electron configuration as follows:

$$\delta_i^v = \frac{Z^v - h}{Z - Z^v - 1} \qquad (2.4.2)$$

where Z^v is the number of valence electrons in atom i, Z the atomic number of atom i, and h the number of hydrogen atoms bound to atom i [23].

Delta values are used to calculate molecular connectivity indices (MCIs). MCIs are based on numerical quantities assigned to substructures. The substructure quantity is calculated as $(\delta_i \delta_j \delta_k \cdots \delta_p)^{-0.5}$ or $(\delta_i^v \delta_j^v \delta_k^v \cdots \delta_p^v)^{-0.5}$, where the multiplication of delta values is carried over all atoms that belong to the particular substructure. Four different substructure types have been applied: path (P), cluster (C), path–cluster (PC), and chain (CH). The number of bonds present in a given substructure is called the *order* of the substructure. A specific MCI is defined as the sum of all substructure quantities considering a substructure of a particular type and a particular order. The

MCI symbol is χ, specified by a left-side superscript representing the order, by a right-side superscript v if the MCI is based on δ^v values, and by a right-side subscript representing the substructure type. If no subscript is indicated, type P is assumed. For example, the MCI $^1\chi$, defined in eq. 2.3.12, is derived from δ values. It is a P-type index of order 1. The analogous valence-corrected MCI, $^1\chi^v$, is defined as

$$^1\chi^v = \sum (\delta_i^v \delta_j^v)^{-0.5} \tag{2.4.3}$$

Subsequent definitions are given for δ-derived MCIs. Corresponding valence-corrected MCI are derived by substituting δ^v for δ.

Path-Type MCIs A path in a molecular graph is an ordered set of consecutive bonds $(b_1, b_2, b_3, \ldots, b_m)$ with the property that bond b_k $(1 \le k \le m)$ starts from the atom where bond b_{k-1} ends. The length of such a path is m. The order of a path equals its length m. The general formula for a path-type MCI is

$$^m\chi = \sum (\delta_i \delta_j \delta_k \cdots \delta_p)^{-0.5} \tag{2.4.4}$$

where δ_i and δ_j are the delta values for the atom pair connected by b_1, δ_j and δ_k are the delta values for the atom pair connected by b_2, and so on. The summation is done over all paths of order m in the molecule. The case $m = 1$ has been considered in definitions 2.3.12 and 2.4.3. The case $m = 0$ specifies the zero-order index, $^0\chi$:

$$^0\chi = \sum (\delta_i)^{-0.5} \tag{2.4.5}$$

Cluster and Path–Cluster MCIs A cluster is a starlike substructure with one central atom and three or more neighbor atoms to which it is connected. The order of a cluster equals the degree $d_z = (\mathbf{v})_z$ of the central atom. The general MCI formula is

$$^{d_z}\chi_{\mathrm{C}} = \sum (\delta_z \delta_{n1} \delta_{n2} \cdots \delta_{ndz})^{-0.5} \tag{2.4.6}$$

where d_z is the delta value of the central atom and δ_{n1}, $\delta_{n2}, \ldots, \delta_{ndz}$ are the delta values of the δ_z neighbor atoms. The summation is done over all clusters of order d_z in the molecule. For example, two third-order clusters are present in 2,5-dimethyl-hexane, one in 2-methylheptane, and none in n-octane. Each fourth-order cluster contains four third-order clusters. Thus, 2,2-dimethylhexane has four third-order clusters and one fourth-order cluster. A path–cluster subgraph is a path that includes at least one additional atom connected to a nonterminal atom of the path. G-II is the simplest path–cluster. It is of fourth order.

$$\begin{array}{c} m \\ | \\ i \diagdown_j \diagup^k \diagdown_l \end{array}$$

G-II

The corresponding MCI is calculated as

$$^4\chi_{\mathrm{PC}} = \sum (\delta_i \delta_j \delta_k \delta_l \delta_m)^{-0.5} \tag{2.4.7}$$

Chain-Type MCIs   A chain is a closed path, a path that ends at the same atom from which it started. CH-type MCIs describe the type of rings that are present in a molecule and the substitution pattern on those rings. For example, an eighth-order chain-type MCI, $^8\chi_{CH}$, is defined for any substructure that corresponds to the graph of either cyclooctane, methylcycloheptane, or ethylcyclohexane. Calculation of CH-type MCIs is performed in analogy to the MCIs defined previously.

Autocorrelation of Topological Structure   Moreau and Broto [24,25] have suggested the autocorrelation vector of a molecular graph as the source for molecular descriptors. This method assumes that each atom i in the graph is uniquely associated with a numeric quantity, q_i, such as the atomic number, atomic mass, $(\mathbf{v})_i$, $(\mathbf{d_S})_i$, δ^v, or electronegativity. The intrinsic atom values of the electrotopological state [26] and the atomic R_D and log K_{ow} parameters [27,28] are other potential atomic descriptors suitable to construct autocorrelation vectors. Generally, the kth element of the autocorrelation vector is defined as

$$(\mathbf{a})_k = \sum q_i q_j \qquad (2.4.8)$$

where the summation is carried over all atom pairs with $k = (\mathbf{D})_{ij} = 0$ including $k = (\mathbf{D})_{ii}$. The dimension of the autocorrelation vector $\mathbf{a}$ is $d_{max} + 1$. The auto-correlation method has been applied, for example, in connection with DARC concentric fragments [29].

General a_N Index   The general a_N-index (GAI) applies to molecules that contain heteroatoms, multiple bonds, and *cis/trans*-specified double-bond locations [30]. This approach is based on the orbital interaction graph of linked atoms (OIGLA) and the orbital interaction matrix of linked atoms (OIMLA). The GAI is computed as the absolute value of the determinant of OIMLA:

$$\mathrm{GAI} = |\det(\mathrm{OIMLA})| \qquad (2.4.9)$$

GAI calculation has been demonstrated for dimethyl methylphosphonate and has been applied to quantitative structure–density correlations for organophosphorus compounds [30]. For an introduction to other quantum-chemical indices, the review of Balasubramanian [31] should be consulted. *Ab initio* descriptors used in the GIPF approach (see Section 1.5) have been discussed by Murray, et al. [32].

Physicochemical Properties as Computable Molecular Descriptors   Any physicochemical property can be a computable molecular descriptor if there exists an algorithm for this property solely from structure input. For example, $(\log K_{ow})_{GCM-n}$ is the logarithm of the n-octanol/water partition coefficient calculated with group contribution model n. Here it is not important how good or bad model n performs in estimating observable K_{ow}. What is important is that model n produces log K_{ow} in a unique way for all training set compounds to assure equally unique calculation of log K_{ow} for new compounds. GCMs for properties such as V_m, R_D, parachor, T_b, or K_{ow}, discussed in subsequent chapters, are appropriate choices to define computable molecular descriptors.

REFERENCES

1. Harary, F., *Proof Techniques in Graph Theory*, 1969. San Drego, CA: Academic Press.
2. Trinajstić, N., *Chemical Graph Theory*, 1983. Boca Raton, FL: CRC Press.
3. Basak, S. C., G. J. Niemi, and G. D. Veith, Optimal Characterization of Structure for Prediction of Properties. *J. Math. Chem.*, 1990: **4**, 185–205.
4. Brezonik, P. L., *Chemical Kinetics and Process Dynamics in Aquatic Systems*, 1994. Boca Raton, FL: CRC Press.
5. Trinajstić, N., et al., *Computational Chemical Graph Theory: Characterization, Enumeration and Generation of Chemical Structures by Computer Methods*, 1991. New York: Ellis Horwood.
6. Mihalic, Z., S. Nikolić, and N. Trinajstić, Comparative Study of Molecular Descriptors Derived from the Distance Matrix. *J. Chem. Inf. Comput. Sci.*, 1992: **32**, 28–37.
7. Mihalic, Z., and N. Trinajstić, Graph-Theoretical Approach to Structure–Property Relationships. *J. Chem. Educ.*, 1992: **69**, 701–712.
8. Thangavel, P., and P. Venuvanalingam, Algorithm for the Computation of Molecular Distance Matrix and Distance Polynomial of Chemical Graphs on Parallel Computers. *J. Chem. Inf. Comput. Sci.*, 1993: **33**, 412–414.
9. Müller, W. R., et al., An Algorithm for Construction of the Molecular Distance Matrix. *J. Comput. Chem.*, 1987: **8**, 170–173.
10. Bersohn, M. A., A Fast Algorithm for Calculation of the Distance Matrix of a Molecule. *J. Comput. Chem.*, 1983: **4**, 110–113.
11. Schultz, H. P., Topological Organic Chemistry: 1. Graph Theory and Topological Indices of Alkanes. *J. Chem. Inf. Comput. Sci.*, 1989: **29**, 227–228.
12. Balaban, A. T., and M. V. Diudea, Real Number Vertex Invariants: Regressive Distance Sums and Related Topological Indices. *J. Chem. Inf. Comput. Sci.*, 1993: **33** 421–428.
13. Estrada, E., Edge Adjacency Relationships and a Novel Topological Index Related to Molecular Volume. *J. Chem. Inf. Comput. Sci.*, 1995: **35**, 31–33.
14. Diudea, M. V., O. Minailiuc, and A. T. Balaban, Regressive Vertex Degree (New Graph Invariants) and Derived Topological Indices. *J. Comput. Chem.*, 1991: **12**, 527–535.
15. Lukovits, I., General Formulas for the Wiener Index. *J. Chem. Inf. Comput. Sci.*, 1991: **31**, 503–507.
16. Gálvez, J., et al., Charge Indexes. New Topological Descriptors. *J. Chem. Inf. Comput. Sci.*, 1994: **34**, 520–525.
17. Bonchev, D., and N. Trinajstić, Information Theory, Distance Matrix, and Molecular Branching. *J. Chem. Phys.*, 1977: **67**, 4517–4533.
18. Bertz, S. H., The First General Index of Molecular Complexity. *J. Am. Chem. Soc.*, 1981: **103**, 3599–3601.
19. Basak, S. C., et al., Determining Structural Similarity of Chemicals Using Graph-Theoretical Indices. *Discrete Appl. Math.*, 1988: **19**, 17–44.
20. Schultz, H. P., and E. B. Schultz, Topological Organic Chemistry: 2. Graph Theory, Matrix Determinants and Eigenvalues, and Topological Indices of Alkanes. *J. Chem. Inf. Comput. Sci.*, 1990: **30**, 27–29.
21. Knop, J. V., et al., On the Determinant of the Adjacency-Plus-Distance Matrix as the Topological Index for Characterizing Alkanes. *J. Chem. Inf. Comput. Sci.*, 1991: **31**, 83–84.
22. Randić, M., P. J. Hansen, and P. C. Jurs, Search for Useful Graph Theoretical Invariants of Molecular Structure. *J. Chem. Inf. Comput. Sci.*, 1988: **28**, 60–68.

23. Sabljic, A., and D. Horvatic, GRAPH III: A Computer Program for Calculating Molecular Connectivity Indices on Microcomputers. *J. Chem. Inf. Comput. Sci.*, 1993: **33**, 292–295.

24. Moreau, G., and P. Broto, Autocorrelation of Molecular Structures: Application to SAR Studies. *Nouv. J. Chim.*, 1980: **4**, 757–764.

25. Moreau, G., and P. Broto, The Autocorrelation of a Topological Structure: A New Molecular Descriptor. *Nouv. J. Chim.*, 1980: **4**, 359–360.

26. Hall, L. H., B. Mohney, and L. B. Kier, The Electrotopological State: Structure Information at the Atomic Level for Molecular Graphs. *J. Chem. Inf. Comput. Sci.*, 1991: **31**, 76–82.

27. Ghose, A. K., A. Pritchett, and G. M. Crippen, Atomic Physicochemical Parameters for Three Dimensional Structure Directed Quantitative Structure–Activity Relationships: III. Modeling Hydrophobic Interactions. *J. Comput. Chem.*, 1988: **9**, 80–90.

28. Viswanadhan, V. N., et al., Atomic Physicochemical Parameters for Three Dimensional Structure Directed Quantitative Structure–Activity Relationships: IV. Additional Parameters for Hydrophobic and Dispersive Interaction and Their Application for an Automated Superposition of Certain Naturally Occurring Nucleoside Antibiotics. *J. Chem. Inf. Comput. Sci.*, 1989: **29**, 163–172.

29. Dubois, J.-É., M. Loukianoff, and C. Mercier, Topology and the Quest for Structural Knowledge. *J. Chim. Phys.*, 1992: **89**, 1493–1506.

30. Xu, L., H.-Y. Wang, and Q. Su, A Newly Proposed Molecular Topological Index for the Discrimination of Cis/Trans Isomers and for the Studies of QSAR/QSPR. *Comput. Chem.*, 1992: **16**, 187–194.

31. Balasubramanian, K., Integration of Graph Theory and Quantum Chemistry for Structure–Activity Relationships. *SAR QSAR Environ. Res.*, 1994: **2**, 59–77.

32. Murray, J. S., et al., Statistically-based Interaction Indices Derived from Molecular Surface Electrostatic Potentials: A General Interaction Properties Function (GIPF). *J. Mol. Struct. (Theochem.)*, 1994: **307**, 55–64.

CHAPTER 3

DENSITY AND MOLAR VOLUME

3.1 DEFINITIONS AND APPLICATIONS

Density is defined as the concentration of matter, measured by the mass per unit volume [1]. The molar volume, V_M, is defined as the volume occupied by 1 mol of a substance. The molar volume of an ideal gas is $22.4140 \, dm^3 \, mol^{-1}$ ($22.4140 \, liter \, mol^{-1}$) at 1 atm pressure and $0°C$. Vapor densities ρ_v are derived through rearrangement of the ideal gas law equation as

$$\rho_v = \frac{PM}{RT} \tag{3.1.1}$$

where ρ_v is the pressure in atm, M the molecular mass in $g \, mol^{-1}$, R the ideal gas constant [$0.082 \, atm \, dm^3 \, (mol \, K)^{-1}$], and T the temperature in K [2]. Liquid densities, ρ_L, can be expressed as the ratio of molecular mass, M, to molar volume V_M:

$$\rho_L = \frac{M}{V_M} \tag{3.1.2}$$

Water attains its maximum density of $0.999973 \, g \, cm^{-3}$ at $3.98°C$ [3]. Liquid organic compounds exhibit densities that are lower or higher than the density of water, usually in the range 0.6 to $3.0 \, g \, cm^{-3}$.

Frequently, the specific gravity, d_{t2}^{t1}, is used, which is defined as the ratio of the weight of any volume of a given liquid at temperature t_1 to the weight of an equal volume of a standard, usually water, at temperature t_2. The commonly reported specific gravity of liquids, d_4^t, in which the standard is water at $4°C$ with its density nearly identical to $1.0000 \, g \, cm^{-3}$, is numerically equal to the density ρ_L at temperature t.

Estimation methods for V_m are needed because some compounds, such as polychlorinated biphenyls and trace organic components of biological materials or environmental matrices are not available in pure form. The V_m value of a compound is related to the energy needed to form a cavity in a solvent during the solubilization process and thus a predictor variable for aqueous solubility [4].

USES FOR DENSITY AND MOLAR VOLUME DATA

- To decide if an immiscible compound floats in water or sinks to the bottom
- To calculate the molar refraction with the Lorentz–Lorenz eq. 4.1.1
- To calculate the parachor with eq. 5.1.1
- To convert kinematic into dynamic viscosity and vice versa (eq. 6.1.1)
- To estimate the liquid viscosity with eq. 6.3.2
- To estimate the liquid viscosity
- To derive cohesion parameters [5]
- To estimate solubilities [6]
- To estimate the n-octanol / water partition coefficient (eq. 13.2.4)

3.2 RELATIONSHIPS BETWEEN ISOMERS

The molar volume, V_M^{25}, of a branched isomer can be expressed as a function of the molar volume of the corresponding normal isomer and certain molecular descriptors. Early systematic studies of this kind include those done by Platt [7] and Greenshields and Rossini [8] and studies referenced therein. The following equation has been derived to relate V_M^{25} of a branched alkane molecule to V_M^{25} of the corresponding n-alkane:

$$V_M^{25} \text{ (branched)} = V_M^{25} \text{ (normal)} + 0.439\,N_{Ct} + 0.578\,N_{Cq} - 1.993\,\Delta P_{C3C}$$
$$- \frac{4.410\,\Delta W}{N_C^2 - N_C} - 3.68 P_4' \tag{3.2.1}$$

where N_{Ct}, N_{Cq}, and P_4' correspond to the branched isomer and ΔP_{C3C} is the difference P_{C3C} (branched) $- P_{C3C}$ (normal). The descriptors are defined in Sections 2.1 and G.2. Equation 3.2.1 is based on 104 hydrocarbons in the range C_5 to C_{30} [8]. A corresponding equation for alkanols has been derived with 39 compounds in the range C_3 to C_7:

$$V_M^{25} \text{ (branched)} = V_M^{25} \text{ (normal)} + 0.493\,N_{OHs} - 0.192\,N_{OHt} + 0.196\,\Delta P_{C3O}$$
$$- 1.374\,\Delta P_{C3C} - \frac{5.209\,\Delta W}{N_C^2 - N_C} \tag{3.2.2}$$

Obama et al. [9] have studied V_M values for isomeric dialkyl ethers. Comparing n-propoxy with i-propoxy alkanes, they derived the following rule:

If a n-propyl group in a dialkyl ether is replaced by an *iso*-propyl V_M group, then V_M increases about $2\,cm^3\,mol^{-1}$, regardless of the nature of the other alkyl groups.   (R-3.2.1)

Obama et al. [9] reported two factors significant in modeling structure–V_M relationships in isomeric dialkyl ethers: (1) the existence of side chains and (2) the difference between the numbers of C atoms in the two alkyl groups. The first factor causes V_M to increase (i.e., a branched dialkyl ether has a greater V_M value than that of its linear isomer):

$$V_M \text{ (branched di-}n\text{-alkyl ether)} > V_M \text{ (unbranched dialkyl ether)} \qquad \text{(R-3.2.2)}$$

The second factor can be accounted for by the following qualitative rule:

$$V_M \text{ (}sym\text{-di-}n\text{-alkyl ether)} > V_M \text{ (}unsym\text{-di-}n\text{-alkyl ether)} \qquad \text{(R-3.2.3)}$$

Ayers and Agruss [10] observed the following relation for dialkyl sulfides (C_6–C_{10}):

$$d_4^t \text{ (di-}n\text{-alkyl sulfide)} > d_4^t \text{ (di-}iso\text{-alkyl sulfide)} \qquad \text{(R-3.2.4)}$$

for t values of 0, 20, and 25°C.

3.3 STRUCTURE–DENSITY AND STRUCTURE–MOLAR VOLUME RELATIONSHIPS

3.3.1 Homologous Series

The observation that the CH_2 group contributes a constant amount to the property of a compound has been made for many properties and also applies to V_M. Van Krevelen [11] has listed the CH_2 contribution to V_M from 12 different GCMs ranging between 16.1 and 16.6 cm^3 mol^{-1}. Jannelli et al. [12] have reported a simple linear relationship between V_M at 25°C and N_{CH_2} for n-alkanenitriles (C_2–C_8). Further, they review analogous relationships for alkanes, alcohols, diols, ethers, and amines.

However, a close inspection of V_M data, especially with respect to a large range in N_{CH_2}, reveals significant deviations from CH_2 constancy. Various studies report the nonlinearity in V_M/N_{CH_2} correlations. Kurtz et al. [13] discuss the nonlinear dependence of V_M on N_{CH_2} and N_{CF_2} for acyclic and cyclic hydrocarbons and perfluorohydrocarbons, respectively. Huggins [14] has reported the following equation for n-alkanes (C_5–C_{18}):

$$V_M^{20} = 27.20 + 16.50 N_{CH_2} + \frac{27}{N_{CH_2}} \qquad (3.3.1)$$

For 1-substituted n-alkanes with the general formula $C_m H_{2m+1} X$, Huggins [15] evaluated the following relationship:

$$V_M^{20} = A + 16.50m + \frac{B}{m} \qquad (3.3.2)$$

where A and B are empirically derived constants characteristic of the substituent X but independent of the chain length m. Constants A and B are shown in Table 3.3.1 for various substituents X.

TABLE 3.3.1   Constants for Eq. 3.3.2 for $C_mH_{2m+1}X$ Compounds [16,17]

$-X$	A	B
$-H$	27.20	27
$-CH_3$	43.9	22
$-F$	29.4	10.6
$-Cl$	38.2	1.0
$-Br$	41.7	-0.5
$-I$	48.9	-3.5
$-OH$	26.1	-2.0
$-SH$	41.65	-1.5
$-C\equiv N$	38.8	-2.8
$-NH_2$	33.4	-2
$-NO_2$	41.7	-6
$-O-N=O$	50.1	3
$-C(=O)OH$	43.3	-2.5
$-C(=O)H$	41.2	-3.2

Huggins [18] also applied eq. 3.3.2 to n-alkyl n-alkanoates with the general formula $C_qH_{2q+1}C(=O)OC_pH_{2p+1}$ and $m = q + p$. The corresponding constants A and B are listed in Table 3.3.2.

TABLE 3.3.2   Constants for Eq. 2.3.2 for n-Alkyl n-alkanoates [16,17]

Series	q	p	A	B
n-Alkyl methanoates	1	>1	32.90	-4.4
n-Alkyl ethanoates	2	>1	33.80	-4.0
n-Alkyl n-propionates	3	>1	34.00	-3.6
Methyl methanoates, ethanoates, n-propionates	1, 2, 3	1	30.90	-2.0
methyl n-alkanoates	>4	1	32.80	-6.7

Source: Refs. 16 and 17.

Smittenberg and Mulder [16,17] introduced the concept of the *limit of a property*, P_∞, into structure–density relationships. P_∞ represents the property of a hypothetical compound with an infinite number of carbon atoms. This concept is especially useful to estimate properties of polymers. For the density at 20°C, P_∞ becomes d_∞^{20}. Smittenberg and Mulder derived the following equation:

$$d^{20} = d_\infty^{20} + \frac{k}{N_C + z} \qquad (3.3.3)$$

where k and z are empirical constants, characteristic for a homologous series. Parameters for this equation are given in Table 3.3.3 for 1-alkanes, 1-alkenes, 1-cyclopentylalkanes, 1-cyclohexylalkanes, and 1-phenylalkanes. For the two

TABLE 3.3.3 Constants for Eq. 3.3.1 [16,17]

Homologous Series	d_∞^{20}	k	z
1-Alkanes	0.8513	-1.3100	0.82
1-Alkenes	0.8513	-1.1465	0.44
1-Cyclopentylalkanes	0.8513	-0.5984	0.00
1-Cyclohexylalkanes	0.8513	-0.5248	0.00
1-Phenylalkanes	0.8513	-0.0535	-4.00

Source: Refs. 16 and 17.

cycloalkyalkane series, the z value was assumed to be zero prior to the analytical derivation of the constant k.

Li et al. [19] expressed V_M^{25} with the following equation:

$$V_M^{25} = V_0^{25} + a_v N_C + \frac{b_v}{N_C - 1} + \frac{c_v}{(N_C - 1)^2} \qquad (3.3.4)$$

where V_0^{25}, a_v, b_v, and c_v are empirical constants. The values of b_v and c_v are given in Table 3.3.4 for various homologous series. Parameter a_v equals 16.4841 cm^3 mol^{-1} for all series.

TABLE 3.3.4 Constants for Eq. 3.3.4 [19]

Homologous Series	N_{compa} [a]	V_0^{25} [b]	b_v [b]	c_v [b]
n-Alkanes	12	45.82233	14.56329	-4.56336
1-Alkenes	12	57.08054	10.37057	-5.33246
n-Alkyl cyclopentanes	3	95.80176	-0.74372	1.64148
n-Alkyl cyclohexanes	3	110.53675	-0.81676	1.02295
n-Alkyl benzenes	3	91.99335	-5.03136	4.71845
1-Alkanethiols	6	59.61365	-5.09148	5.03934
2-Alkanethiols	5	77.53475	-4.61859	4.64630
1-Alkanols	7	43.56824	-3.74475	3.36719
2-Alkanols	5	61.65620	-6.88659	7.21123
n-Alkanoic acids	6	43.84111	-1.17385	-1.09678

[a] Number of compounds used to derive parameters.
[b] In cm^3 mol^{-1}.
Source: Reprinted with permission from Ref. 19. Copyright (1955) American Chemical Society.

3.3.2 Molecular Descriptors

Correlations of Kier and Hall Kier and Hall [20] have studied relationships between d_4^{20} and various MCIs. For example, they report the following equation for alkanes (C$_5$–C$_9$):

$$d_4^{20} = 0.7955 - \frac{0.4288}{^1\chi} + 0.0131\,^3\chi_P - 0.079\,^5\chi_P + 0.051\,^4\chi_{PC}$$
$$+ 0.0035\,^5\chi_{PC} \qquad n = 46, \quad s = 0.0024, \quad r = 0.9971 \qquad (3.3.5)$$

where n is the sample size, s the standard deviation, and r the correlation coefficient. The descriptor $^1\chi$ is highly correlated with V_M of n-alkanes; hence its reciprocal, $1/^1\chi$, has been taken in this correlation to account for the contribution of the CH_2 chain to d_4^{20}. The higher-order descriptors $^3\chi_P$, $^5\chi_P$, $^4\chi_{PC}$, and $^5\chi_{PC}$ are indicative for the various classes of methyl- and ethyl-substituted alkanes. Similar correlation have been given for alkanols, aliphatic ethers, and aliphatic acids.

Correlations of Needham, Wei, and Seybold   Similar to the correlation of Kier and Hall, the correlation [21] uses MCIs as independent variables. The model has been derived for alkanes (C_2–C_9):

$$V_M(cm^3\,mol^{-1}) = 22.9(\pm 0.7) + 14.6(\pm 0.2)^0\chi + 16.2(\pm 0.4)^1\chi - 7.3(\pm 0.2)^3\chi_P$$
$$- 4.5(\pm 0.3)^4\chi_P + 2.1(\pm 0.3)^5\chi_c$$
$$n = 69, \quad s = 0.5, \quad r^2 = 0.999, \quad F = 14294 \qquad (3.3.6)$$

where V_M is at 20 °C and F is Fisher's significance factor.

Correlation of Estrada   Estrada [22] derived the following simple, linear correlation between $V_M^{20}(cm^3\,mol^{-1})$ and the ε index for alkanes (C_5–C_9) (eq. 2.3.6):

$$V_M^{20} = 57.8501 + 30.8559\epsilon \qquad n = 69, \quad s = 2.032, \quad r = 0.9931, \quad F = 4831$$
$$(3.3.7)$$

Correlations of Bhattacharjee, Basak, and Dasgupta   Bhattacharjee et al. [23] found the following relationship for V_M^{20} of haloethanes:

$$V_M^{20} = 26.2053 + 22.9394V_g \qquad n = 27, \quad s = 5.5091, \quad r^2 = 0.9215 \qquad (3.3.8)$$

where V_g is the geometric volume. Bhattacharjee and Dasgupta [24] gave the following equation for V_M^{20} of alkanes (C_1–C_8):

$$V_M^{20} = 51.8165 + 13.9583V_g \qquad n = 37, \quad s = 4.73, \quad r^2 = 0.9991 \qquad (3.3.9)$$

Correlation of Grigoras   Grigoras [25] derived a simple linear correlation to estimate $V_M^{20}(cm^3\,mol^{-1})$ for liquid compounds, including saturated, unsaturated, and aromatic hydrocarbons, alcohols, acids, esters, amines, and nitriles:

$$V_M^{20} = -19.9 + 0.684A \qquad n = 137, \quad s = 6.6, \quad r = 0.970, \quad F = 2207 \qquad (3.3.10)$$

where A is the total molecular surface area based on contact atomic radii [25].

Correlation of Xu, Wang, and Su   Xu et al. [26] have studied correlations between d_4^{20} and the general a_N index, GAI. For dialkyl methylphosphonates, $CH_3P(=O)(OR)_2$, they reported the following relationship:

$$d_4^{20} = 0.85 + 0.23GAI \qquad n = 14, \quad s = 0.006, \quad r = 0.97 \qquad (3.3.11)$$

3.4 GROUP CONTRIBUTION APPROACH

Scaled Volume Method of Girolami Girolami [27] has suggested a simple atom contribution method that allows density estimation with an accuracy of $0.1\,\mathrm{g\,cm^{-3}}$ for a variety of liquids. The methods include correction factors for certain hydrogen-bonding groups and fused rings. The atom contributions are atomic volumes with values relative to the atomic volume of hydrogen, which has been set to 1. The contribution associated with elements belonging to the first, second, or third row of the periodic table are listed in Table 3.4.1. The *scaled volume*, V_{scal}, of a molecule is calculated as the sum of the atom contributions of its constituent atoms. Then the density ρ is given as follows:

$$\rho = M(5\,V_{\mathrm{scal}})^{-1} \tag{3.4.1}$$

where M is the molar mass and the factor 5 allows the density to be expressed in units of $\mathrm{g\,cm^{-3}}$. The temperature has not been specified. Densities calculated with formula 3.4.1 have to be increased by 10% for each of the following groups:

- Hydroxyl group (–OH)
- Carboxylic acid group [–C(=O)OH]
- Primary or secondary amino group (–NH_2, –NH–)
- Amide group [–C(=O)NH_2 and *N*-substituted derivatives]
- Sulfoxide group [–S(=O)–]
- Unfused ring

For a system of fused rings, a 7.5% increase has been recommended for each ring. Based on 166 test liquids, the correlation between observed and estimated densities has been analyzed by a least square fit ($\rho_{\mathrm{obs}} = 1.01\rho_{\mathrm{est}} - 0.006$; $r^2 = 0.982$). For only two compounds (acetonitrile and dibromochloromethane) does the error exceed $0.1\,\mathrm{g\,cm^{-3}}$. Girolami has demonstrated this method for dimethylethylphosphine,

TABLE 3.4.1 Atom Contributions to V_{scal} in Girolami's Method [27]

Element	Atom Contribution
H	1
Short period	
Li to F	2
Na to Cl	4
Long period	
K to Br	5
Rb to I	7.5
Cs to Bi	9

Source: Reprinted with permission from the *Journal of Chemical Education*, Vol. 71, No. 11, 1994, pp. 962–964; copyright © 1994, Division of Chemical Education, Inc.

cyclohexanol, ethylenediamine, sulfolane, and 1-bromonaphthalene. Although more accurate estimation methods are available, this method is unique with respect to its simplicity and its broad applicability range including organic, inorganic, and metal–organic liquids.

Method of Horvath   The atom contribution method of Horvath has been reported to estimate V_M^{25} for halogenated hydrocarbons and ethers in the range C_1 to C_4 [28, p. 314]:

$$V_M^{25} = 7.7 + 8.2N_H + 13.4N_F + 22.3N_{Cl} + 24.7N_{Br} \qquad (3.4.2)$$

where V_M^{25} is the molar volume at 25°C and N_H, N_F, N_{Cl}, and N_{Br} are the number of hydrogen, fluorine, chlorine, and bromine atoms per molecule, respectively.

Method of Schroeder   Schroeder's method has been evaluated for the molar volume at the normal boiling point, V_b [29]. The contributions, including extra bond and ring contributions, are shown in Table 3.4.2. The equation is

$$V_b = \sum n_i (V_b)_i + \sum n_i^{extra} (V_b)_i^{extra} \qquad (3.4.3)$$

where $(V_b)_i$ and $(V_b)_i^{extra}$ are the corresponding contributions, and n_i and n_i^{extra}, respectively, count the number of their occurrences per molecule.

TABLE 3.4.2 V_b Contributions in Schroeder's Method [28]

Atom	$(V_b)_i{}^a$	Atom	$(V_b)_i{}^a$	Ring / Bond	$(V_b)_i^{extra\ a}$
H	7	F	10.5	Ring	−7
C	7	Cl	24.5	Single bond	0
N	7	Br	31.5	Double bond	7
O	7	I	38.5	Triple bond	14
S	21				

a $(V_b)_i$ in $cm^3 (g\,mol)^{-1}$.
Source: Reprinted with permission from Ref. 29. Copyright (1997) McGraw-Hill Book Company.

GCM Values for V_M at 20 and 25°C.   Most GCM values for V_M are based on group definitions that are more discriminative than those used in the models above. Examples are the method of Exner [29] to estimate V_M at 20°C and the method of Fedors [31] to estimate V_M at 25°C. GCM values have been reviewed by van Krevelen [11]. Highly discriminative methods have been developed by Dubois and Loukianoff [32] and by Constantinou et al. [33]. These methods are discussed briefly below.

LOGIC Method   The *local-to-global-information-construction* (LOGIC) *method* [32] has been applied to the estimation of the density of alkanes at 25°C. In this method the groups are atom-centered substructures, FREL$_B$ (fragment reduced to an environment that is limited). A FREL is an atomic group characterized by its vertex degree and the vertex degree of its neighbor atoms in the first and second neighbor sphere. FRELs are denoted as four-digit integers encoding the neighbor sphere information. The contributions for C_{sp3} atoms are listed in Table 3.4.3.

TABLE 3.4.3   FREL Contributions to V_M^{25} in cm^3 mol^{-1}

Code	Value	Code	Value	Code	Value
V1000	0.000	V3222	2.272	V4311	6.257
V1100	20.483	V3300	10.917	V4320	4.458
V1110	16.788	V3310	5.311	V4321	0.459
V1111	14.791	V3311	1.920	V4322	−4.543
V2000	40.946	V3320	−0.304	V4330	−1.710
V2100	28.640	V3321	−3.831	V4331	−6.341
V2110	24.084	V3322	−7.359	V4332	−10.972
V2111	21.341	V3330	−5.919	V4333	−15.603
V2200	16.334	V3331	−9.446	V4400	3.716
V2210	11.780	V3332	−12.974	V4410	−2.713
V2211	9.256	V3333	−16.501	V4411	−7.344
V2220	7.284	V4000	59.166	V4420	−9.143
V2221	4.571	V4100	45.819	V4421	−13.774
V2222	−0.976	V4110	39.285	V4422	−18.405
V3000	50.364	V4111	34.243	V4430	−15.572
V3100	37.222	V4200	31.395	V4431	−20.203
V3110	31.616	V4210	24.968	V4432	−24.834
V3111	27.834	V4211	20.380	V4433	−29.465
V3200	24.023	V4220	18.492	V4440	−22.002
V3210	18.393	V4221	14.071	V4441	−26.633
V3211	14.929	V4222	6.896	V4442	−31.264
V3220	12.963	V4300	17.091	V4443	−35.895
V3221	9.489	V4310	11.149	V4444	−40.526

Source: Reprinted with permission from Ref. 32. Copyright (1993) Gordon and Breach Publishers, World Trade Center, Lausanne, Switzerland.

Application of the LOGIC method is demonstrated in Figure 3.4.1 with 3,8-diethyldecane. The estimated density is 0.7740 g cm^{-3}. Experimental values of 0.7770 and 0.7340 g cm^{-3} are known at 20 and 80°C, respectively [34]. Interpolation yields a value of 0.7732 g cm^{-3} at 25°C, which compares favorably with the estimated value.

Method of Constantinou, Gani, and O'Connell   The approach of Constantinou et al. [33] has been described for T_b in Section 9.3. The analog model for V_M of liquids at 25°C is

$$V_M - 0.01211\,\text{m}^3\,\text{kmol}^{-1} = \sum n_i (V_{M1})_i + W \sum m_j (V_{M2})_j \qquad n = 312$$

$W = 0: \quad s = 0.00236\,\text{m}^3\,\text{kmol}^{-1}, \quad \text{AAE} = 0.00139\,\text{m}^3\,\text{kmol}^{-1}, \quad \text{AAPE} = 1.16\%$

$W = 1: \quad s = 0.00192\,\text{m}^3\,\text{kmol}^{-1}, \quad \text{AAE} = 0.00105\,\text{m}^3\,\text{kmol}^{-1}, \quad \text{AAPE} = 0.89\%$

(3.4.4)

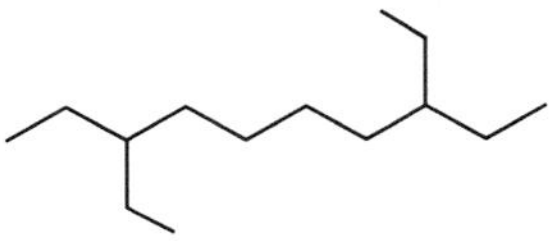

3,8-Diethyldecane

1. Molar volume at 25°C

V1110	4(20.483)	81.932
V2110	4(24.084)	96.336
V2200	2(16.334)	32.668
V2210	2(11.780)	23.560
V3300	2(10.917)	21.834

$$V_M = 256.330 \, \text{cm}^3 \, \text{mol}^{-1}$$

2. Molar mass: $M = 198.40 \, \text{g mol}$

$$\rho = 0.7740 \, \text{g cm}^{-3} \text{ at } 25°C$$

Figure 3.4.1   Estimation of ρ at 25°C for 3,8-diethyldecane using the LOGIC method.

where $(V_{M1})_i$ is the contribution of the first-order group type i which occurs n_i times in the molecule, and $(V_{M2})_j$ is the contribution of the second-order type j with m_j occurrences in the molecule. W is zero or 1 for a first- or second-order approximation, respectively, and the statistical parameters are $s = [\sum (T_{b,\text{fit}} - T_{b,\text{obs}})^2 / n]^{1/2}$, $\text{AAE} = (1/n) \sum |T_{b,\text{fit}} - T_{b,\text{obs}}|$, and $\text{AAPE} = (1/n) \sum |T_{b,\text{fit}} - T_{b,\text{obs}}| / T_{b,\text{obs}} \, 100\%$.

3.5   TEMPERATURE DEPENDENCE

Vapor and liquid densities decrease with increasing temperature. Here, the following temperature coefficient of density is considered:

$$\frac{d\rho}{dt} = \frac{\rho^{t_1} - \rho^{t_2}}{t_2 - t_1} \tag{3.5.1}$$

where $t_2 < t_1$. The $d\rho/dt$ term can be assumed to be approximately constant between 0 and 40°C, unless a phase change occurs within this range [3]; $d\rho/dt$ depends strongly on the molecular structure. Table 3.5.1 compares temperature coefficients for various temperature intervals of some structurally different compounds.

Densities of various hydrocarbon compounds have been reported at 20, 25, and 30°C [35–39] including temperature coefficients of density at 25°C. For 1-alkenes, for example, the coefficients decrease with increasing N_C in the range from $0.001034 \, \text{g cm}^{-3} \, °\text{C}^{-1}$ for 1-pentene to $0.000733 \, \text{g cm}^{-3} \, °\text{C}^{-1}$ for 1-dodecene. For hydrocarbons and various compounds containing heteroatoms, density/temperature correlations have been presented as a polynomial function of the type

$$\rho(T) = a_0 + a_1 T + a_2 T^2 + a_3 T^3 \tag{3.5.2}$$

TABLE 3.5.1 Temperature Coefficients for Densities of Selected Compounds

ρ^{t_1} (g cm^{-3})	t_1(°C)	ρ^{t_2} (g cm^{-3})	t_2(°C)	$d\rho/dt$ (g cm^{-3} °C^{-1})
1,1,3-Trimethylcyclopentane [35]				
0.74825	20.0	0.74392	25.0	0.00087
0.74392	25.0	0.73958	30.0	0.00087
2-Bromobiphenyl [40]				
1.4192	0.0	1.4081	9.8	0.00113
1.3973	20.0	1.3875	30.0	0.00098
1.3875	30.0	1.3771	40.0	0.00104
Halothane (2-bromo-2-chloro-1,1,1-trifluoroethane) [41]				
1.8721	18.0	1.8703	19.5	0.00120
1.8690	20.0	1.8646	21.4	0.00314
1.8606	23.0	1.8521	25.5	0.00340
1.8482	27.6	1.840.2	30.4	0.00286
1.8308	33.8	1.8231	36.4	0.00296
1.8172	39.5	1.8146	40.5	0.00260
Diethanolamine [42]				
1.4750	20.0	1.4735	25.0	0.00030
1.4735	25.0	1.4721	30.0	0.00028
2-Methyltetrahydrofuran [43]				
0.8662	3.1	0.8621	6.9	0.00108
0.8559	13.7	0.8514	17.7	0.00113
0.8449	24.4	0.8418	25.9	0.00187
0.8360	33.6	0.8332	38.0	0.00064

where a_0, a_1, a_2, and a_3 are empirical, compound-specific coefficients. In Tables A.1 through A.6 in Appendix A, the coefficients, the applicable temperature range, and references are listed for various compounds. Note that the temperature range does not always include the range of environmental interest. Extrapolation to the environmental range should be performed only with great care and with critical consideration of eventual conclusions drawn using those extrapolated values, especially with respect to possible phase transition.

Rutherford [43] reports the use of the modified Rackett equation to correlate the density for alkyl chlorides (1-chloroethane, 1-chloropropane, and 1-chlorobutane) and bromides (bromomethane, bromoethane, and bromopropane) as a function of temperature, pressure, and critical point data.

Method of Grain Grain has proposed a method to estimate the liquid density from normal boiling point data using the following equation [2]:

$$\rho_L = MV_b^{-1}\left(3 - 2\frac{T_x}{T_b}\right)^n \tag{3.5.3}$$

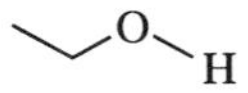

$$\overset{\diagdown\diagup O\diagdown}{}\text{H}$$
Ethanol

1. Classification: alcohol$\rightarrow n = 0.25$
2. Boiling point: $T_b = 351.5\,\text{K}$ [45]
3. Molar mass: $M = 46.069\,\text{g mol}^{-1}$
4. Schroeder's V_b:

i	n_i	A_i	$n_i A_i$
H	6	7	42
C	2	7	14
O	1	7	7

$$V_b = 63\,\text{cm}^3(\text{g mol})^{-1}$$

5. With eq. 3.5.3: $\rho_L = 46.069(63^{-1})\left[3-2\left(\dfrac{283.15}{351.5}\right)\right]^{0.25}$

$$\rho_L = 0.7938\,\text{g cm}^{-3}\ \text{at}\ 10°\text{C}$$

Figure 3.5.1   Estimation of ρ_L at 10 °C for ethanol using Grain's method.

where ρ_L is the density in g cm^{-3}, M the molecular mass in g mol^{-1}, V_b the molar volume at the boiling point in $\text{cm}^3\,(\text{g·mol})^{-1}$, T_x and T_b are in K, and n is 0.25 for alcohols, 0.29 for hydrocarbons, and 0.31 for other organic compounds. This method is provided in the Toolkit. V_b is estimated with Schroeder's method (Table 3.4.2). Grain's method is demonstrated in Figure 3.5.1 by estimating the density of ethanol at 10°C. The corresponding experimental value is $0.79789\,\text{g cm}^{-3}$ [44].

Fisher [46] reports a relationship for n-alkanes of essentially any chain length that allows estimation of d_4 as a function of temperature and of N_C:

$$d_4(t) = \frac{-0.29796 + 0.86555N_C - (0.0025302 + 0.00057592N_C)t}{N_C + 1.215} \tag{3.5.4}$$

where t is in °C. The relationships has been derived from data for n-$C_{36}H_{74}$ and lower homologs. To give an example, eq. 3.5.4 has been applied to estimate the density of n-tetranonacontane (n-$C_{94}H_{190}$) at 115 and at 135°C:

$$d_4(115\,°\text{C}) = 0.7829 \qquad d_4(135\,°\text{C}) = 0.7710$$

Experimental values of $0.7833\,\text{g mL}^{-1}$ at 115 °C and of $0.7714\,\text{g mL}^{-1}$ at 135 °C reported by Reinhard and Dixon [47] are in good agreement with the estimated values.

REFERENCES

1. Lide, D. R., and H. P. R. Frederikse, *CRC Handbook of Chemistry and Physics*, 75th ed. (1994–1995), 1994. Boca Raton, FL: CRC Press.

2. Nelken, L. H., Densities of Vapors, Liquids and Solids, in *Handbook of Chemical Property Estimation*, W. J. Lyman, W. F. Reehl, and D. H. Rosenblatt, Editors, 1990. Washington, DC: American Chemical Society.

3. Riddick, J. A., *Organic Solvents: Physical Properties and Methods of Purification*, 4th ed., 1986. New York: Wiley.

4. Kamlet, M. J., et al., Linear Solvation Energy Relationships: 36. Molecular Properties Governing Solubilities of Organic Nonelectrolytes in Water. *J. Pharma. Sci.*, 1986: **75**, 338–349.

5. Barton, A. F. M., *CRC Handbook of Solubility and Other Cohesion Parameters*, 2nd ed., 1991. Boca Raton, FL: CRC Press.

6. McAuliffe, C., Solubility in Water of Paraffin, Cycloparaffin, Olefin, Acetylene, Cyclo-olefin, and Aromatic Hydrocarbons. *J. Phys. Chem.*, 1966: **70**, 1267–1275.

7. Platt, J. R., Prediction of Isomeric Differences in Paraffin Properties. *J. Phys. Chem.*, 1952: **56**, 328–336.

8. Greenshields, J. B., and F. D. Rossini, Molecular Structure and Properties of Hydrocarbons and Related Compounds. *J. Phys. Chem.*, 1958: **62**, 271–280.

9. Obama, M., et al., Densities, Molar Volumes, and Cubic Expansion Coefficients of 78 Aliphatic Ethers. *J. Chem. Eng. Data*, 1985: **30**, 1–5.

10. Ayers, G. W. J. and M. S. Agruss, Organic Sulfides: Specific Gravities and Refractive Indices of a Number of Aliphatic Sulfides. *J. Am. Chem. Soc.*, 1939: **61**, 83–85.

11. Van Krevelen, D. W., *Properties of Polymers*, 3rd ed., 1990. Amsterdam: Elsevier.

12. Janelli, L., M. Pansini, and R. Jalenti, Partial Molar Volumes of C2–C6 *n*-Alkanenitriles and Octanenitrile in Dilute Aqueous Solutions at 298.16 K. *J. Chem. Eng. Data*, 1984: **29**, 266–269.

13. Kurtz, S. S., Jr., et al., Molecular Increment of Free Volume in Hydrocarbons, Fluoro-hydrocarbons, and Perfluorocarbons. *J. Chem. Eng. Data*, 1962: **7**, 196–202.

14. Huggins, M. L., Densities and Refractive Indices of Liquid Paraffin Hydrocarbons. *J. Am. Chem. Soc.*, 1941: **63**, 116–120.

15. Huggins, M. L., Densities and Refractive Indices of Unsaturated Hydrocarbons. *J. Am. Chem. Soc.*, 1941: **63**, 916–920.

16. Smittenberg, J., and D. Mulder, Relation Between Refraction, Density and Structure of Series of Homologous Hydrocarbons: I. Empirical Formulae for Refraction and Density at 20°C of *n*-Alkanes and *n*-α-Alkenes. *Recueil*, 1948: **67**, 813–825.

17. Smittenberg, J., and D. Mulder, Relation Between Refraction, Density and Structure of Series of Homologous Hydrocarbons: II. Refraction and Density at 20°C of *n*-Alkyl-cyclopentanes, -cyclohexanes and -benzenes. *Recueil*, 1948: **67**, 826–838.

18. Huggins, M. L., Densities and Optical Properties of Organic Compounds in the Liquid State: V. The Densities of Esters from Fatty Acids and Normal Alkohols. *J. Am. Chem. Soc.*, 1954: **76**, 847–850.

19. Li, K., et al., Correlation of Physical Properties of Normal Alkyl Series of Compounds. *J. Phys. Chem.*, 1955: **60**, 1400–1406.

20. Kier, L. B., and L. H. Hall, *Molecular Connectivity in Chemistry and Drug Research*, 1976. San Diego, CA: Academic Press.

21. Needham, D. E., I.-C. Wei, and P. G. Seybold, Molecular Modeling of the Physical Properties of the Alkanes. *J. Am. Chem. Soc.*, 1988: **110**, 4186–4194.

22. Estrada, E., Edge Adjacency Relationships and a Novel Topological Index Related to Molecular Volume. *J. Chem. Inf. Comput. Sci.*, 1995: **35**, 31–33.

23. Bhattacharjee, S., and P. Dasgupta, Molecular Property Correlation in Haloethanes with Geometric Volume. *Comput. Chem.*, 1992: **16**, 223–228.

24. Bhattacharjee, S., and P. Dasgupta, Molecular Property Correlation in Alkanes with Geometric Volume. *Comput. Chem.*, 1994: **18**, 61–71.

25. Grigoras, S., A Structural Approach to Calculate Physical Properties of Pure Organic Substances: The Critical Temperature, Critical Volume and Related Properties. *J. Comput. Chem.*, 1990: **11**, 493–510.

26. Xu, L., H.-Y. Wang, and Q. Su, A Newly Proposed Molecular Topological Index for the Discrimination of Cis/Trans Isomers and for the Studies of QSAR / QSPR. *Comput. Chem.*, 1992: **16**, 187–194.

27. Girolami, G. S., A Simple "Back of the Envelope" Method for Estimating the Densities and Molecular Volumes of Liquids and Solids. *J. Chem. Educ.*, 1994: **71**, 962–964.

28. Horvath, A. L., *Molecular Design. Chemical Structure Generation from the Properties of Pure Organic Compounds*, 1992, Amsterdam: Elsevier B.V.

29. Reid, R. C., J. M. Prausnitz, and T. K. Sherwood *The Properties of Gases and Liquids*, 1977, 3rd ed., New York, N4: Mcgraw-Hill Book Company.

30. Exner, O., Additive Physical Properties: II. Molar Volume as an Additive Property. *Collect. Czech. Chem. Commun.*, 1967: **32**, 1–22.

31. Fedors, R. F., A Method for Estimating Both the Solubility Parameters and Molar Volumes of Liquids. *Poly. Eng. Sci.*, 1974: **14**, 147–154.

32. Dubois, J. E., and M. Loukianoff, DARC "Logic Method" for Molal Volume Prediction. *SAR QSAR Environ. Res.*, 1993: **1**, 63–75.

33. Constantinou, L., R. Gani, and J. P. O'Connell, Estimation of the Acentric Factor and the Liquid Molar Volume at 298 K Using a New Group Contribution Method. *Fluid Phase Equilibria*, 1995: **103**, 11–22.

34. Korosi, G., and E. S. Kovats, Density and Surface Tension of 83 Organic Liquids. *J. Chem. Eng. Data*, 1981: **26**, 323–332.

35. Forziati, A. F., and F. D. Rossini, Physical Properties of Sixty API– NBS Hydrocarbons. *J. Res. Nat. Bur. Stand.*, 1949: **43**, 473–476.

36. Camin, D. L., A. F. Forziati, and F. D. Rossini, Physical Properties of n-Hexadecane, n-Decylcylopentane, n-Decylcyclohexane, 1-Hexadecene and n-Decylbenzene. *J. Phys. Chem.*, 1954: **58**, 440–442.

37. Camin, D. L., and F. D. Rossini, Physical Properties of 14 American Petroleum Institute Research Hydrocarbons, C9 to C15. *J. Phys. Chem.*, 1955: **59**, 1173–1179.

38. Camin, D. L., and F. D. Rossini, Physical Properties of the 17 Isomeric Hexenes of the API Research Series. *J. Phys. Chem.*, 1956: **60**, 1446–1451.

39. Camin, D. L., and F. D. Rossini, Physical Properties of 16 Selected C7 and C8 Alkene Hydrocarbons. *J. Phys. Chem.*, 1960: **5**, 368–372.

40. Amey, L., and A. P. Nelson, Densities and Viscosities of 2-Bromobiphenyl and 2-Iodobiphenyl. *J. Chem. Eng. Data*, 1982: **27**, 253–254.

41. Francesconi, R., F. Comelli, and D. Giacomini, Excess Enthalpy for the Binary System 1,3-Dioxolane + Halothane. *J. Chem. Eng. Data*, 1990: **35**, 190–191.

42. Murrieta-Guevara, F., and A. T. Rodriguez, Liquid Densities as a Function of Temperature of Five Organic Solvents. *J. Chem. Eng. Data*, 1984: **29**, 204–206.

43. Shinsaka, K., N. Gee, and G. R. Freeman, Densities Against Temperature of 17 Organic Liquids and of Solid 2,2-Dimethylpropane. *J. Chem. Thermodyn.*, 1985: **17**, 1111–1119.

44. Rutherford, W. M., Viscosity and Density of Some Lower Alkyl Chlorides and Bromides. *J. Chem. Eng. Data*, 1988: **33**, 234–237.

45. Schroeder, M. R., B. E. Poling, and D. B. Manley, Ethanol Densities Between −50 and 20°C. *J. Chem. Eng. Data*, 1982: **27**, 256–258.

46. Merck, *The Merck Index: An Encyclopedia of Chemicals, Drugs, and Biologicals*, 11th. ed., 1989. Rahway, NJ: Merck & Co., Inc.

47. Fisher, C. H., How to Predict *n*-Alkane Densities? These Equations Predict Density for C20 and Heavier Alkanes. *Chem. Eng.*, 1989: **96**(10), 195.

48. Reinhard, R. R., and J. A. Dixon, Tetranonacontane. *J. Org. Chem.*, 1965: **30**, 1450–1453.

CHAPTER 4

REFRACTIVE INDEX AND MOLAR REFRACTION

4.1 DEFINITIONS AND APPLICATIONS

The refractive index of a medium or a compound, n, is defined as c/v, the ratio of the velocity of light in vacuum (c) to the velocity of light in the medium or compound (v). Reported values usually refer to the ratio of the velocity in air to that in the air-saturated compound [1]. If the light of the sodium D line (wavelength $l = 589.3\,nm$) is used at temperature $t\,(°C)$, the measured refractive index is denoted n_D^t.

The Lorentz–Lorenz equation [2] defines the molar refraction, R_D, as a function of the refractive index, density, and molar mass:

$$R_D = \left(\frac{n_D^2 - 1}{n_D^2 + 2}\right)\frac{M}{\rho} \tag{4.1.1}$$

Some authors use the notation R_M or R_m instead of R_D. This notation, however, can be confused with the use of R_M, the chromatographic retention index. Here R_D is used to indicate molar refraction, where the subscript D refers to the sodium D line used for measurement.

The reverse calculation of n_D^t from R_D is given by the following equation:

$$n_D = \left(\frac{M + 2\rho R_D}{M - \rho R_D}\right)^{1/2} \tag{4.1.2}$$

from which n_D is obtained as the positive value of the square root. Refractive indices, n_D^{20}, for various liquid compounds, in most cases along with d_4^{20} values, can be found in the *CRC Handbook of Chemistry and Physics* [3] or in the *Merck Index* [4].

USES FOR REFRACTIVE INDEX AND MOLAR REFRACTION DATA

- To assess purity of compound
- To calculate the molecular electronic polarizability $\alpha_E = 3P_D/(4\pi)$ [5]
- To estimate the boiling point with Meissner's method [6]
- To estimate liquid viscosity [7]

4.2 RELATIONSHIPS BETWEEN ISOMERS

Greenshields and Rossini [8] derived equations for the molar refraction in analogy to eqs. 3.2.1 and 3.2.2. The following equation has been given to relate R_D^{25} of an alkane molecule to R_D^{25} of the corresponding n-alkane:

$$R_D^{25}(\text{branched}) = R_D^{25}(\text{normal}) + 0.017N_{Ct} + 0.047N_{Cq} - 0.121\Delta P_{C3C} \quad (4.2.1)$$

In this equation N_{Ct} and N_{Cq} correspond to the branched isomer and ΔP_{C3C} is the difference P_{C3C} (branched) $- P_{C3C}$ (normal). Equation 4.2.1 has been derived with 66 compounds in the range C_5 to C_{10}. For alkanols, Greenshields and Rossini [8] derived an equation based on 26 compounds in the range C_3 to C_6:

$$R_D^{25}(\text{branched}) = R_D^{25}(\text{normal}) - 0.026\,N_{OHs} - 0.116N_{OHt} + 0.018\Delta P_{C3O}$$
$$+ 0.017N_{Ct} + 0.047N_{Cq} - 0.121\Delta P_{C3C} \quad (4.2.2)$$

where ΔP_{C3O} is the difference P_{C3O} (branched) $- P_{C3O}$ (normal).

Ayers and Agruss [9] observed the following relation for dialkyl sulfides (C_6-C_{10}) for t equal to 20 and 25 °C:

$n_D^t(\text{di-}n\text{-alkyl sulfide}) > n_D^t(\text{di-}iso\text{-alkyl sulfide})$

$\Delta R_D[\text{RE} : |\text{CCC}, \text{C(C)C}|] = 0.07\,\text{cm}^{-3} \qquad (\text{R} - 4.2.1)$

(note that both n-alkyl groups have to be exchanged)

With data for 1-methoxy ethyl ketones, $R-C(=O)-CH(CH3)-O-CH3$ (C_7-C_9) [10], the following rule was developed:

$\Delta R_D^{20}[\text{RE} : |\text{CCC}, \text{C(C)C}|] = 0.05\,\text{cm}^{-3} \qquad (\text{R} - 4.2.2)$

4.3 STRUCTURE–R_D RELATIONSHIPS

Molar refractivity depends on the number of electrons in a molecule that can interact with through-passing light. The more atoms a molecule has (i.e., the larger the molecular size), the higher the number of electrons, and thus the stronger through-passing light is bent. Below, correlations between molar refractivity and molecular structure will be considered in some detail. The following qualitative rule should be kept in mind as a general guideline:

> Generally, the number of electrons increases with increasing size of the molecule, and, thus, the ability of the molecule to bend light.   (R-4.3.1)
> Rule of thumb: Molar refractivity increases with increasing molecular size.

Kurtz and co-workers [11], for example, discuss the relation between R_D of hydrocarbons and molecular descriptors such as the number of carbon atoms, N_C, in the molecule, the number of chain and ring carbons, the number of side chains, and the number of double bonds.

Method of Smittenberg and Mulder   In analogy to eq. 3.3.3, Smittenberg and Mulder [12,13] evaluated the following equation for alkanes, 1-alkenes, 1-cyclo-pentylalkanes, 1-cyclohexylalkanes, and 1-phenylalkanes:

$$n_D^{20} = n_{D,\infty}^{20} + \frac{k}{N_C + z} \qquad (4.3.1)$$

where $n_{D,\infty}^{20}$ is the refraction index at 20°C for $N_C = \infty$ and k and z are empirical constants, characteristic for the series. The parameters of this equation for the specified compound classes are given in Table 4.3.1.

TABLE 4.3.1   Constants for Eq. 4.3.1

Homologous Series	n_D^{20}, ∞	k	z
n-Alkanes	1.47519	0.68335	0.816
1-Alkenes	1.47500	-0.55506	0.374
1-Cyclopentylalkanes	1.4752	-0.3920	0
1-Cyclohexylalkanes	1.4752	-0.3438	0
1-Phenylalkanes	1.4752	0.1125	-2.30

Source: Compiled from Refs. 12 and 13.

Method of Li et al.   Li et al. [14] modeled n_D^{25} for various homologous series with the following equation:

$$R_D^{25} = R_0^{25} + a_R N_C \qquad (4.3.2)$$

where R_0^{25} is an empirical constant and a_R equals 4.64187 cm^3 mol^{-1} for all series. The derived R_0^{25} are given in Table 4.3.2.

Van der Waals Volume–Molar Refraction Relationships.   Bhatnagar et al. [15] have found a significant correlation between R_D and V_{vdW} for alkyl halides

TABLE 4.3.2   Constants for Eq. 4.3.2

Homologous Series	N_{comp}^{a}	$R_0^{25\,b}$
n-Alkanes	12	6.72066
1-Alkenes	12	10.93704
n-Alkyl cyclopentanes	3	23.14251
n-Alkyl cyclohexanes	3	27.76551
n-Alkyl benzenes	3	26.55060
1-Alkanethiols	6	14.50273
2-Alkanethiols	5	19.17105
1-Alkanols	7	8.23182
2-Alkanols	4	12.85626
n-Alkanoic acids	6	8.26657

[a] Number of compounds used to derive parameters.
[b] In $cm^3\,mol^{-1}$.

Source: Reprinted with permission from Ref. 14. Copyright (1956) American Chemical Society.

(C_2–C_5, F, Cl, Br, I), alkanols (C_4–C_9), monoalkyl amines (C_3–C_{11}), and dialkyl amines (C_3–C_{10}):

$$R_D = -4.713 + 26.613 V_{vdW} \quad n = 65, \quad s = 2.832, \quad r = 0.915, \quad F_{1,63} = 324.81$$

$$(4.3.3)$$

and for monosubstituted phenols:

$$R_D = -2.912 + 33.427 V_{vdW} \quad n = 25, \quad s = 2.539, \quad r = 0.934, \quad F_{1,23} = 156.12$$

$$(4.3.4)$$

The phenol substituents include alkyl (C_1–C_4), alkoxy (C_1–C_5), and a few other groups. However, the effect of the ring position on R_D was not evaluated.

Geometric Volume–Molar Refraction Relationships   Similar to the van der Waals volume–molar refraction approach, Bhattacharjee and Dasgupta [16] studied correlations between R_D and the geometric volume for alkanes and haloalkanes. For alkanes (C_1–C_8), the following equation has been reported:

$$R_D^{20} = 4.2923 + 4.4887 V_g \quad n = 35, \quad s = 40.50, \quad r^2 = 0.9923 \quad (4.3.5)$$

where V_g is the geometric volume. For haloalkanes, the appplicable equation depends on the particular pattern of halogen substitution [17].

Correlations of Kier and Hall   Kier and Hall [5] found the following relationship between R_D and MCIs for alkanes (C_5–C_{10}):

$$R_D^{20} = 4.008 + 7.331\,^1\chi + 2.423\,^2\chi_P + 0.454\,^3\chi_P - 0.619\,^4\chi_C$$
$$- 0.141\,^4\chi_{PC} \quad n = 46, \quad s = 0.027, \quad r = 0.9999$$

$$(4.3.6)$$

Similarly, they derived relationships for alkenes, alkylbenzenes, alkanols, dialkyl ethers, mono-, di-, and trialkyl amines, and alkyl halides. For example, the equation for dialkyl ethers (C_4–C_8) is

$$R_D^{20} = 3.569 + 9.070 \ ^1\chi^v + 1.953 \ ^3\chi_C \qquad n = 9, \quad s = 0.291, \quad r = 0.9989$$

$$(4.3.7)$$

Correlations of Needham, Wei, and Seybold   Similar to the correlation of Kier and Hall, the correlation of Needham et al. [18] uses MCIs as independent variables. The model has been derived for alkanes (C_2–C_9):

$$R_D^{20}(cm^3 = -0.8(\pm 0.1) + 3.8(\pm 0.02) \ ^0\chi + 4.6(\pm 0.1) \ ^1\chi - 0.98(\pm 0.03) \ ^3\chi_p$$
$$- 0.63(\pm 0.04) \ ^4\chi_p - 0.25(\pm 0.06) \ ^5\chi_p$$
$$n = 69, \quad s = 0.05, \quad r^2 = 0.9999, \quad F = 152558 \qquad (4.3.8)$$

4.4   GROUP CONTRIBUTION APPROACH FOR R_D

Method of Ghose and Crippen   The method of Ghose and Crippen [19] uses 120 different atom types. They are described for the corresponding K_{ow} model in Chapter 12. A training set of 538 compounds was employed. Observed versus calculated R_D showed a correlation coefficient of 0.998 and a standard deviation of

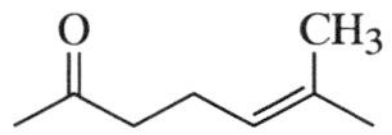

6-Methyl-5-heptane-2-one

Atomic group	Contribution terms	
: CH_3R (No. 1)	3(2.9680)	8.9040
$H[C_{sp^3}]$ (No. 46)	6(0.8447)	5.0682
$H[\alpha\text{-}C]$ (No. 51)	3(0.8188)	2.4564
$=CR_2$ (No. 17)	1(3.9392)	3.9392
$=CHR$ (No. 16)	1(4.2654)	4.2654
$H[C_{sp^2}]$ (No. 47)	1(0.8939)	0.8939
: CH_2R_2 (No. 1)	2(2.9116)	5.8232
$H[C_{sp^3}]$ (No. 46)	2(0.8447)	1.6894
$H[\alpha\text{-}C]$ (No. 51)	2(0.8188)	1.6376
: $Al\text{-}C(=X)\text{-}Al$ (No. 38)	1(3.9031)	3.9031
$=O$ (No. 58)	1(1.4429)	1.4429
	$R_D^{20} =$	40.0233 cm^3

Using eq. 4.1.2 with $M = 126.19$ g mol^{-1} and $\rho^{20} = 0.8508$ g cm^{-3} [20]: $n_D^{20} = 1.4522$

Figure 4.4.1   Estimation of R_D^{20} and n_D^{20} for 6-methyl-5-heptene-2-one using the method of Ghose and Crippen [21].

0.774. Comparison of predicted R_D values with the values observed for a set of 82 test compounds gave a correlation coefficient of 0.996 and a standard deviation of 1.553. An example for the application of this method is shown for 6-methyl-5-heptene-2-one in Figure 4.4.1. An experimental n_D^{20} value of 1.4404 is known [20]. This method has been integrated into the Toolkit.

4.5 TEMPERATURE DEPENDENCE OF REFRACTIVE INDEX

The temperature dependence of the refractive index has been evaluated with the empirical Eykman equation:

$$C_{Eyk} = \frac{(n_D^t)^2 - 1}{[(n_D^t)^2 + 0.4]\rho^t} \qquad (4.5.1)$$

where C_{Eyk} is a temperature-independent constant [22]. For example, Gibson and Kincaid [23] have reported experimentally derived C_{Eyk} values of 0.7506, 0.7507, and 0.7504 cm^3 g^{-1} at 25, 35, and 45°C, respectively. Kurtz et al. [24] demonstrated the applicability of eq. 4.5.1 for different temperature ranges with data on liquid hydrocarbons, alkanols, alkanoic acids, and their esters, phenols, and phenol-alkanones.

Smith and Kiess [25] reported an average change of -0.00043 in n_D per degree over the range 0 to 30°C derived with the three trimethylethylbenzenes shown in Figure 4.5.1.

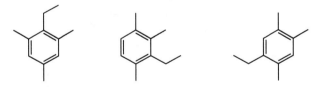

Ethylmesitylen   3-Ethylpseudocumene   5-Ethylpseudocumene

Figure 4.5.1 Molecular structure of ethylmesitylene, 3-ethylpseudocumene, and 5-ethyl-pseudocumene.

REFERENCES

1. Riddick, J. A., *Organic Solvents: Physical Properties and Methods of Purification*, 4th ed., 1986. New York: Wiley.
2. Nelken, L. H., Index of Refraction, in *Handbook of Chemical Property Estimation*, W. J. Lyman, W. F. Reehl, and D. H. Rosenblatt, Editors, 1990. Washington, DC: American Chemical Society, p. 26.
3. Lide, D. R. and H. P. R. Frederikse, *CRC Handbook of Chemistry and Physics*, 75th ed. (1994–1995), 1994. Boca Raton, FL: CRC Press.
4. Merck, *The Merck Index: An Encyclopedia of Chemicals, Drugs, and Biologicals*, 11th ed., 1989. Rahway, NJ: Merck & Co., Inc.

5. Kier, L. B. and L. H. Hall, *Molecular Connectivity in Chemistry and Drug Research*, 1976. San Diego, CA: Academic Press.

6. Rechsteiner, C. E., Boiling Point, in *Handbook of Chemical Property Estimation*, W. J. Lyman, W. F. Reehl, and D. H. Rosenblatt, Editors, 1990. Washington, DC: American Chemical Society.

7. Lagemann, R. T., A Relation Between Viscosity and Refractive Index. *J. Am. Chem. Soc.*, 1945: **67**, 498–499.

8. Greenshields, J. B., and F. D. Rossini, Molecular Structure and Properties of Hydrocarbons and Related Compounds. *J. Phys. Chem.*, 1958: **62**, 271–280.

9. Ayers, G. W. J., and M. S. Agruss, Organic Sulfides: Specific Gravities and Refractive Indices of a Number of Aliphatic Sulfides. *J. Am. Chem. Soc.*, 1939: **61**, 83–85.

10. Wallace, W. P., and H. R. Henze, Keto Ethers: X. 1-Methoxyethyl Alkyl Ketones. *J. Am. Chem. Soc.*, 1942: **64**, 2882–2883.

11. Kurtz, S. S., Jr., et al., Molecular Increment of Free Volume in Hydrocarbons, Fluorohydrocarbons, and Perfluorocarbons. *J. Chem. Eng. Data*, 1962: **7**, 196–202.

12. Smittenberg, J., and D. Mulder, Relation Between Refraction, Density and Structure of Series of Homologous Hydrocarbons: I. Empirical Formulae for Refraction and Density at 20°C of *n*-Alkanes and *n*-α-Alkenes. *Recueil*, 1948: **67**, 813–825.

13. Smittenberg, J., and D. Mulder, Relation Between Refraction, Density and Structure of Series of Homologous Hydrocarbons: II. Refraction and Density at 20°C of *n*-Alkylcyclopentanes, -cyclohexanes and -benzenes. *Recueil*, 1948: **67**, 826–838.

14. Li, K., et al., Correlation of Physical Properties of Normal Alkyl Series of Compounds. *J. Phys. Chem.*, 1956: **60**, 1400–1406.

15. Bhatnagar, R. P., P. Singh, and S. P. Gupta, Correlation of van der Waals Volume with Boiling Point, Solubility and Molar Refraction. *Indian J. Chem.*, 1980: **19B**, 780–783.

16. Bhattacharjee, S., and P. Dasgupta, Molecular Property Correlation in Alkanes with Geometric Volume. *Comput. Chem.*, 1994: **18**, 61–71.

17. Bhattacharjee, S., and P. Dasgupta, Molecular Property Correlation in Haloethanes with Geometric Volume. *Comput. Chem.*, 1992: **16**, 223–228.

18. Needham, D. E., I.-C. Wei, and P. G. Seybold, Molecular Modeling of the Physical Properties of the Alkanes. *J. Am. Chem. Soc.*, 1988: **110**, 4186–4194.

19. Viswanadhan, V. N., et al., An Estimation of the Atomic Contribution to Octanol–Water Partition Coefficient and Molar Refractivity from Fundamental Atomic and Structural Properties: Its Uses in Computer Aided Drug Design. *Math. Comput. Model.*, 1990: **14**, 505–510.

20. Baglay, A. K., L. L. Gurariy, and G. G. Kuleshov, Physical Properties of Compounds Used in Vitamin Synthesis. *J. Chem. Eng. Data*, 1988: **33**, 512–513.

21. Ghose, A. K., and G. M. Crippen, Atomic Physicochemical Parameters for Three Dimensional Structure Directed Quantitative Structure–Activity Relationships I. *J. Comput. Chem.*, 1986: **7**; 565–577.

22. Dreisbach, R. R., Applicability of the Eykman Equation. *Ind. Eng. Chem.*, 1948: **40**; 2269–2271.

23. Gibson, R. E., and J. F. Kincaid, The Influence of Temperature and Pressure on the Volume and Refractive Index of Benzene. *J. Am Chem. Soc.*, 1938: **60**; 511–518.

24. Kurtz, S. S., Jr., S. Amon, and A. Sankin, Effect of Temperature on Density and Refractive Index. *Ind. Eng. Chem.*, 1950: **42**; 174–176.

25. Smith, L. I., and M. A. Kiess, Polymethylbenzenes: XXIII. The Preparation and Physical Properties of 3- and 5-Ethylpseudocumenes and of Ethylmesitylene. *J. Am. Chem. Soc.*, 1939: **61**, 284–288.

CHAPTER 5

SURFACE TENSION AND PARACHOR

5.1 DEFINITIONS AND APPLICATIONS

Surface Tension   The surface tension of a liquid is defined as the force per unit length exerted in the plane of the liquid's surface [1,2]. Some authors use the symbol σ, others use γ to represent the surface tension. The surface tension is expressed in $dyn\,cm^{-1}$. For most organic liquids, σ is between 25 and 40 $dyn\,cm^{-1}$ at ambient temperatures. The surface tension of water is 72 $dyn\,cm^{-1}$ at 25°C. For polyhydroxy compounds, the surface tension ranges up to 65 $dyn\,cm^{-1}$.

UNIT CONVERSION

$$1\,dyn = 10^5 N$$

$$1\,dyn\,cm^{-1} = 1\,mN\,m^{-1}$$

Parachor   The parachor is defined as follows [1]:

$$\text{parachor} = \sigma^{1/4} M \left(\rho_L - \rho_{vap} \right)^{-1} \tag{5.1.1}$$

where surface tension is $dyn\,cm^{-1}$, M, the molecular mass in $g\,mol^{-1}$, ρ_L, is the liquid density in $g\,cm^{-3}$; and ρ_{vap}, the density of the saturated vapor in $g\,cm^{-3}$.

The *parachor* does not have a readily apparent physicochemical meaning; it is useful as a parameter for estimating a range of other properties, especially those related to liquid–liquid interactions.

> ## USES FOR SURFACE TENSION AND PARACHOR DATA
>
> - To describe emulsification behavior of liquids
> - To calculate interfacial tension between organic liquids and water
> - To describe and model chemical spreading in a spill

5.2 PROPERTY–PROPERTY AND STRUCTURE–PROPERTY RELATIONSHIPS

Multiparametric correlations between σ and physicochemical and molecular properties are known. Needham et al. [2] reported the following model for alkanes $(C_2–C_9)$:

$$\sigma(\mathrm{dyn\,cm^{-1}}) = 28.1(\pm 0.5) - 50.1(\pm 2.2)N_C^{-1} - 1.7(\pm 0.2)T_m + 0.11(\pm 0.02)T_m^2$$
$$+ 0.56(\pm 0.02)P_{C3C} + 0.05(\pm 0.02)T_3$$
$$n = 68, \quad s = 0.2, \quad r^2 = 0.989, \quad F = 1152 \tag{5.2.1}$$

where σ is the surface tension at 20°C and T_m is the melting point in °C. The following model applies for the same set of compounds but employs solely molecular-structure-based descriptors [2]:

$$\sigma(\mathrm{dyn\,cm^{-1}}) = 31.3(\pm 0.7) - 35.6(\pm 1.3)(^1\chi)^{-1} - 0.8(\pm 0.1)^2\chi + 0.8(\pm 0.1)^3\chi_p$$
$$+ 0.4(\pm 0.1)^3\chi + 0.9(\pm 0.1)^5\chi_c$$
$$n = 68, \quad s = 0.2, \quad r^2 = 0.986, \quad F = 845 \tag{5.2.2}$$

Stanton and Jurs [3] developed a model for a more diverse set of compounds, including hydrocarbons, halogenated hydrocarbons, alkanols, ethers, ketones, and esters. The model has been evaluated with 31 compounds, using, among others, charge partial surface area (CPSA) descriptors:

$$\sigma(\mathrm{dyn\,cm^{-1}}) = (12.87 \pm 2.02) + (-29.16 \pm 3.64)\mathrm{FNSA\text{-}2}$$
$$+ (387.62 \pm 39.81)\mathrm{FPSA\text{-}3} + (-1.29 \pm 0.19)\mathrm{RPCS}$$
$$+ (0.28 \pm 0.06)\mathrm{RNCS} + (106.71 \pm 22.82)\,^7\chi_c^v$$
$$+ (-9.51 \pm 2.95)\mathrm{TOPSYM} \quad n = 31, \quad s = 2.32,$$
$$r = 0.953, \quad F = 39.14, \quad F(0.9; 6; 25) = 2.02 \tag{5.2.3}$$

where σ is the surface tension at 20°C and FNSA-2 is the total fractional negative charged surface area (F_{partial}: 64.18), FPSA-3 is the fractional positive atomic charged weighted partial surface area (F_{partial} : 94.79), RPCS is the relative positive charged surface area (F_{partial}: 45.66), RNCS is the relative negative charged surface area (F_{partial}: 23.39), $^7\chi_c^v$ is the seventh order valence-corrected chain molecular connectivity index (F_{partial}: 21.86), and TOPSYM is the topological symmetry (F_{partial}: 10.41) [3]. Within the model, polar interaction information is supplied solely by the CPSA descriptors. The latter also account for the greatest amount of the variance (i.e., they show the largest partial F_{partial} values).

5.3 GROUP CONTRIBUTION APPROACH

Estimation methods for the surface tension of a liquid are based on eq. 5.1.1. Generally, $\rho_{\text{vap}} \ll \rho_L$ and ρ_{vap} can be ignored. Thus one obtains

$$\sigma = (\text{parachor} \cdot M^{-1} \rho_L)^4 \qquad (5.3.1)$$

Estimation of σ with 5.3.1 requires solely the input of ρ_L and parachor. Parachor can be derived from molecular structure with schemes based on group additivity. Exner [4] gives an excellent review and discussion of various group contribution methods for parachor. A very simple method has been developed by McGowan [5] employing only atomic contribution and the number of bonds, N_{bonds}:

$$\text{parachor} = \sum n_i A_i - 19 N_{\text{bonds}} \qquad (5.3.2)$$

where A_i is the contribution for atom of type i and n_i the number of atoms of type i. The summation is done over all atomic types that occur in the molecule. Estimation of surface tension at two different temperatures based on McGowan's parachor is given in Figures 5.3.1 and 5.3.2 for pentanenitrile and 1,2-dimethoxyethane, respectively. Experimental surface tensions are available for comparison: 27.39 dyn cm^{-1} (20°C) for pentanenitrile and 17.71 dyn cm^{-1} (80°C) for 1,2-dimethoxyethane [6]. The GCM of McGowan is available in the Toolkit.

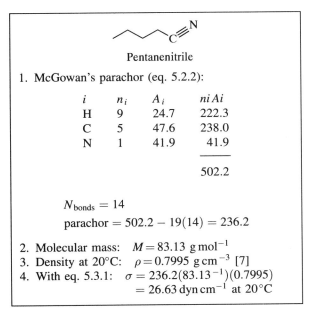

Pentanenitrile

1. McGowan's parachor (eq. 5.2.2):

i	n_i	A_i	$n_i A_i$
H	9	24.7	222.3
C	5	47.6	238.0
N	1	41.9	41.9
			502.2

$N_{\text{bonds}} = 14$

parachor $= 502.2 - 19(14) = 236.2$

2. Molecular mass: $M = 83.13$ g mol^{-1}
3. Density at 20°C: $\rho = 0.7995$ g cm^{-3} [7]
4. With eq. 5.3.1: $\sigma = 236.2(83.13^{-1})(0.7995)$
$= 26.63$ dyn cm^{-1} at 20°C

Figure 5.3.1 Estimation of σ at 20°C for pentanenitrile.

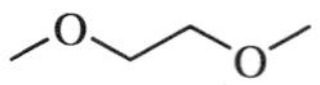

1,2-Dimethoxyethane

1. McGowan's parachor (eq. 5.2.2):

i	n_i	A_i	$n_i A_i$
H	10	24.7	247.0
C	4	47.6	190.4
O	2	36.2	72.4
			509.8

$N_{bonds} = 15$

Parachor $= 509.8 - 19(15) = 224.8$

2. Molecular mass: $M = 90.12$ g mol^{-1}
3. Density at 80°C: $\rho = 0.8092$ g cm^{-3} [7]
4. With eq. 5.3.1: $\sigma = 224.8(90.12^{-1})(0.8092)$
   $= 16.60$ dyn cm^{-1} at 80°C

Figure 5.3.2 Estimation of σ at 80°C for 1,2-dimethoxyethane.

5.4 TEMPERATURE DEPENDENCE OF SURFACE TENSION

The highest value for the surface tension of pure compounds is found at the triple point. Between this and the critical point, the surface tension gradually decreases with rising temperature and becomes zero at the critical point [7]. Jasper [8] has reported linear σ / T correlation for a variety of compounds:

$$\sigma(T) = a_0 + a_1 T \qquad (5.4.1)$$

where a_0 and a_1 are compound-specific constants. Reid et al. [9] and Horvath [7] discuss methods to estimate $\sigma(T)$ that require various properties as input such as the normal boiling point, T_b, the critical temperature, T_c, and the critical pressure, p_c.

Othmer Equation The Othmer equation relates $\sigma(T)$ to the critical temperature, T_c, and a reference point given by σ_{ref} at T_{ref}:

$$\sigma(T) = \sigma_{ref}[(T_c - T)(T_c - T_{ref})]^{11/9} \qquad (5.4.2)$$

Yaws et al. [10] have compiled and evaluated the parameters needed in the Othmer equation for over 600 compounds. An example is shown in Figure 5.4.1, where the surface tension of acetic anhydride at 16.5°C has been estimated.

Temperature Dependence of Parachor The parachor may be considered as being nearly independent of temperature. For example, parachor values at different temperatures have been reported for dimethyl sulfoxide: 182.9 (50°C), 184.7 (100°C), 185.7(150°C), 185.4 (200°C) [11], for diphenyl-p-isopropylphenyl phosphate: 786 (50°C), 791 (100°C), 794 (150°C), 795 (200°C), 793 (240°C) [12], and for

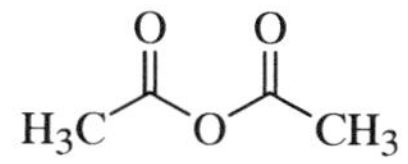

$$H_3C \underset{O}{\overset{O}{\parallel}} \underset{}{\overset{}{}} O \underset{}{\overset{O}{\parallel}} CH_3$$

Acetic anhydride

1. Parameters: $\sigma_{ref} = 32.65$ dyn cm^{-1} at $T_{ref} = 20°C$ [10]
   Range: from $T_f = 199.0$ to $T_c = 569.1$ K
   Units: σ in dyn cm^{-1}, T in K

2. With eq. 5.4.2: $\sigma = 32.65 \left[\left(\dfrac{569.1 - 289.7}{569.1 - 293.2} \right) \right]^{11/9}$

   $\sigma = 33.16$ dyn cm^{-1} at 16.5°C

Figure 5.4.1   Estimation of σ of acetic anhydride at 16.5°C using eq. 5.4.2.

hexamethylenetetramine (urotropin): 315.5 (20°C), 314. 8 (25°C), 314.9 (35°C), 315.4 (45°C) [13]. Owen et al. [14] reported the following result derived with 16 tertiarty alkanols: The parachor increases by 0.2% per 10°C rise in temperature.

REFERENCES

1. Grain, C. F., Interfacial Tension with Water, in *Handbook of Chemical Property Estimation*, W. J. Lyman, W. F. Reehl, and D. H. Rosenblatt, Editors, 1990. Washington, DC: American Chemical Society.

2. Reid, R. C., J. M. Prausnitz, and T. K. Sherwood, *The Properties of Gases and Liquids*. 3rd ed., 1977. New York: McGraw-Hill.

3. Needham, D. E., I.-C. Wei, and P. G. Seybold, Molecular Modeling of the Physical Properties of the Alkanes. *J. Am. Chem. Soc.*, 1988: **110**, 4186–4194.

4. Stanton, D. T., and P. C. Jurs, Development and Use of Charged Partial Surface Area Structural Descriptors for Quantitative Structure–Property Relationship Studies. *Anal. Chem.*, 1990: **62**, 2323–2329.

5. Exner, O., Additive Physical Properties, III. Re-examination of the Additive Character of Parachor. *Collect. Czech. Chem. Commun.*, 1967: **32**, 24–54.

6. Rechsteiner, C. E., Boiling Point, in *Handbook of Chemical Property Estimation*, W. J. Lyman, W. F. Reehl, and D. H. Rosenblatt, Editors, 1990. Washington, DC: American Chemical Society.

7. Korosi, G., and E. S. Kovats, Density and Surface Tension of 83 Organic Liquids. *J. Chem. Eng. Data*, 1981: **26**, 323–332.

8. Horvath, A. L., *Molecular Design: Chemical Structure Generation from the Properties of Pure Organic Compounds*, 1992. Amsterdam: Elsevier.

9. Jasper, J. J., The Surface Tension of Pure Liquid Compounds. *J. Phys. Chem. Ref. Data*, 1972: **1**, 841–1009.

10. Yaws, C. L., H. -C. Yang, and X. Pan, 633 Organic Chemicals: Surface Tension Data. *Chem. Eng.*, 1991: Mar, 140–150.

11. Golubkov, Y. V., et al., Density, Viscosity, and Surface Tension of Dimethyl Sulfoxide. *Zh. Prikl. Khim.*, 1982: **55EE**, 702–703.

12. Tarasov, I. V., Y. V. Golubkov, and R. I. Luchkina, Density, Viscosity, and Surface Tension of Diphenyl-*p*-isopropylphenyl Phosphate. *Zh. Prikl. Khim*, 1982: **56EE**, 2578–2580.

13. Huang, T. C., et al., A Study of the Parachor of Hexamethylenetetramine (Urotropine). *J. Am. Chem. Soc.*, 1938: **60**; 489.

14. Owen, K., O. R. Quayle, and E. M. Beavers, A Study of Organic Parachors: (II). Temperature, and (III) Constitutive Variations of Parachors of a Series of Tertiary Alcohols. *J. Am. Chem. Soc.*, 1939: **61**, 900–905.

CHAPTER 6

DYNAMIC AND KINEMATIC VISCOSITY

6.1 DEFINITIONS AND APPLICATIONS

Viscosity might be described as an internal resistance of a gas or a liquid to flow. Viscosity data are reported as dynamic viscosity, η, or as kinematic viscosities, ν, which are related through density, ρ, by the following equation:

$$\eta = \nu\rho \qquad (6.1.1)$$

where η is presented in units of centipoise (cP), and ν is expressed in units of centistokes (cS).

CONVERSION OF VISCOSITY UNITS

$1 \text{ centipoise} = 10^3 \text{ Pa} \cdot \text{s} = 10^3 \text{N m}^{-2} \cdot \text{s}$

$1 \text{ centistoke} = 10^6 \text{ m}^2 \text{ s}^{-1}$

The viscosity of water at 20°C is 1 cP. For most organic compounds, η is observed in the range 0.3 to 20 cP at environmental temperatures [1].

USES FOR VISCOSITY DATA

- To describe and model transport processes of gases and liquids (fluids)
- To evaluate the pumbability of a liquid
- To assess the spreadability of spills
- To calculate the fluidity which is the reciprocal of the viscosity
- To estimate diffusion coefficients

6.2 PROPERTY–VISCOSITY AND STRUCTURE–VISCOSITY RELATIONSHIPS

Gas viscosity generally decreases with increased molecular size. This trend is reversed for liquids, for which the viscosity increases with increasing N_C within homologous series [2]. The latter observation is confirmed for alkanes, alkanethiols, and n-alkyl β-ethoxypropionates in Tables 6.2.1a-c. Linear correlations between viscosity and N_C have been evaluated, for example, for n-alkyl n-alkoxypropionates [3] and n-alkyl carbonates of methyl and butyl lactates [4,5]. Similarly, correlations

TABLE 6.2.1a Densities and Viscosities of Some n-Alkanes at 20°C

Alkane	ρ (g cm^{-3})	ν (10^6 m^2 s^{-1})
n-Hexane	0.66131	0.4695
n-Heptane	0.68434	0.60013
n-Octane	0.70275	0.76971
n-Decane	0.72995	1.2543
n-Dodecane	0.74946	1.9743
n-Tetradecane	0.76309	3.0189
n-Hexadecane	0.77253	4.4614

Source: Reprinted with permission from Ref. 10. Copyright (1991) American Chemical Society.

TABLE 6.2.1b Densities and Viscosities of Some 1-Alkanethiols at 20°C

Alkane	ρ (g cm^{-3})	η (cP)
Ethanethiol	0.83914	0.293
1-Propanethiol	0.84150	0.399
1-Pentanethiol	0.84209	0.639
1-Hexanethiol	0.84242	0.813
1-Heptanethiol	0.84310	1.043

Source: Compiled from Refs. 11 and 12.

TABLE 6.2.1c Densities and Viscosities of Some n-Alkyl β-Ethoxypropionates at 20°C

n-Alkyl Group	d_4^{20}	η (cP)
Methyl	0.9751	0.180
Ethyl	0.9490	0.260
Propyl	0.9354	0.475
Butyl	0.9256	0.681
Pentyl	0.9191	2.079
Hexyl	0.9120	2.368
Octyl	0.9028	3.437
Decyl	0.8960	4.630

Source: Reprinted with permission from Ref. 8. Copyright (1948) American Chemical Society.

between viscosity and molecular weight have been reported for alkanes, alkyl cylopentanes and alkyl cyclohexanes, 1-alkenes, alkylbenzenes, 1-alkanols, isoalkanols, alkanones, alkanoic acids, and esters [6], covering a molar mass range between 30 and 300 g mol^{-1} and a viscosity range between 0.25 and 10 cP at 20°C. In some series, viscosities reproduced for the first and second member exhibit up to 60% deviation from the experimental value, where an average error of less than 5% has been found for the higher members.

Correlations of viscosity with density and refractive index have been evaluated for various homologous series [7] and correlations between viscosity and boiling point and between viscosity and vapor pressure have been reported, for example, for *n*-alkyl β-ethoxypropionates [8]. Viscosity correlations with vapor pressure are represented by the Porter equation [9]:

$$\log_{10} \eta = a_0 + a_1 p_{vap} \tag{6.2.1}$$

where η and p_{vap} are the viscosity and vapor pressure at the same temperature and a_0 and a_1 are empirical, compound-specific coefficients. Equation 6.2.1 has been studied in combination with the group contribution approach and is described in section 6.3.

6.3 GROUP CONTRIBUTION APPROACHES FOR VISCOSITY

For alkanes, the logarithm of viscosity has been correlated with atomic and with bond contributions to estimate η at 0 and 20°C [13]. Considering a broader range of structural variety, neither the viscosity nor its logarithm is a constitutionally additive property. Application of the group contribution approach is based on additive parameters that allow viscosity estimations in combination with other experimental data such as density or vapor pressure. The *viscosity-constitutional constant*, I_{vc}, is such an additive parameter:

$$I_{vc} = m_\rho M \tag{6.3.1}$$

where M is the molecular weight in g mol^{-1} and m_ρ is the compound-specific viscosity-density constant in cm^3 g^{-1} defined with the following equation [14]:

$$\log_{10}[\log_{10} \eta(mP)] = m_\rho \rho - 2.9 \tag{6.3.2}$$

where ρ is the density in g cm^{-3}. In the temperature range from 0 to 60°C, m_ρ has shown to be temperature independent. Sounders has presented a group contribution scheme to calculate I_{vc} and to estimate η in this temperature range from the corresponding density value [14].

Skubla [9] has designed a group contribution scheme which applies for various homologous series. His method relies on eq. 6.2.1 where both coefficients a_0 and a_1 have to be derived from group contributions and with respect to N_C. The model applies for *n*-alkanes, 1-alkenes, *n*-alkylcyclopentanes and *n*-alkylcyclohexanes, alkylbenzenes, 1-bromoalkanes, 1-alkanols, di-*n*-alkyl ethers, carboxylic acids and esters, 1-alkanethiols, 1-aminoalkanes, dialkylamines, alkaneamides, and some substituted benzenes. Examples how to use the method have been given for butaneamide and *o*-bromotoluene [9].

The method of van Velzen discussed by Grain [1] requires solely molecular structure input. Again, temperature coefficients constitute the additive parameters related with terms for functional groups and various corrections for configurational factors.

Method of Joback and Reid   The Joback and Reid method [15] applies for liquid hydrocarbons, halogenated hydrocarbons, and O-containing compounds:

$$\eta_L(\mathrm{N\,m^{-2}\cdot s}) = M\exp\left[\frac{\sum n_i(\Delta\eta_A)_i - 597.82}{T} + \sum n_i(\Delta\eta_B)_i - 11.202\right] \quad (6.3.3)$$

where M is the molecular mass in $\mathrm{g\,mol^{-1}}$ and T is the temperature in K. The method employs two terms of group contributions, denoted by A and B. For each term the summation is over all group types i. $(\Delta\eta_A)_i$ and $(\Delta\eta_B)_i$ are the contribution to terms A and B, respectively, for the ith group type and n_i is the number of times the group occurs in the molecule. Application of this model to 4-methyl-2-pentanone is demonstrated in Figure 6.3.1 for η_L at 35°C. The estimated value is 0.641 cP, compared to the value of 0.494 cP found in the literature [16].

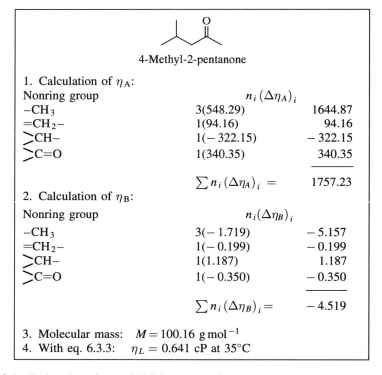

Figure 6.3.1   Estimation of η_L at 35°C for 4-methyl-2-pentanone using the method of Joback and Reid [15].

6.4   TEMPERATURE DEPENDENCE OF VISCOSITY

There is a wide-spread literature on methods for temperature-dependent viscosity estimation. Their discussion and further references can be found elsewhere [1,2,17,18,19,20,21]. Usually, these methods are based on various input data, such as density, boiling point, and critical point. Dynamic viscosities of most gases increase with increasing temperature. Dynamic viscosities of most liquids, including water, decrease rapidly with increasing temperature [18].

6.4.1   Compound-Specific Functions

The Arrhenius equation has been employed to correlate viscosity–temperature data of liquid hydrocarbons:

$$\frac{1}{\eta} = A e^{-E/RT} \tag{6.4.1}$$

where A and E are compound-specific parameters and T is in K [16,19,20]. Bingham [21] has developed equations fitting viscosity–temperature data for hydrocarbons and heterofunctional compounds, including halogenated hydrocarbons, alkanols, alkanoic acids, and esters. Frequently, viscosity–temperature correlations are expressed by the equation

$$\ln \eta(T) = b_0 + \frac{b_2}{T - b_3} \tag{6.4.2}$$

where b_0, b_1, and b_2 are empirical, compound-specific coefficients and T is in K. Examples are listed in Appendix B in Tables B.1 through B.3. Polynomial fitting has also been applied to viscosity–temperature data:

$$\ln \eta(T) = a_0 + a_1 T^{-1} + a_2 T^{-2} + a_3 T^{-3} \tag{6.4.3}$$

with the compound-specific coefficients a_0, a_1, a_2, and a_3 given in Table B.4 for selected hydrocarbons. Yaws et al. [22] used the following equation to present viscosity–temperature data between the melting point and the critical point for structurally diverse compounds with five to seven C atoms:

$$\log_{10}\eta_L(\text{cP}) = A + B T^{-1} + C T + D T^2 \tag{6.4.4}$$

where T is in K. An example is presented in Figure 6.4.1 calculating η_L for 4-methyl-2-pentanone at 35°C. Riggio et al. [16] reported an experimental value of 0.494 cP (compare with the performance of the method of Joback and Reid in Figure 6.3.1).

Method of Cao, Knudsen, Fredenslund, and Rasmussen   The method of Cao et al. [23] is based on a statistical thermodynamic model for pure liquids and liquid mixtures. It requires the input of the compound properties V_M and ΔH_v and two

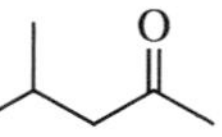

4-Methyl-2-pentanone

1. Temp. coefficient: $A = -3.0570$, $B = 5.0050 \times 10^2$, $C = 6.5038 \times 10^{-3}$,
   $D = -8.8243 \times 10^{-6}$ [22]
   Range: 246–571 K
   Units: η_L in cP, T in K

2. With eq. 6.4.4: $\log_{10} \eta_L = -3.0570 + \dfrac{5.0050 \times 10^2}{308.2} + 6.5038 \times 10^{-3}(308.2)$

$$- 8.8243 \times 10^{-6}(308.2^2)$$
$$= -0.2668$$
$$\eta_L = 0.541 \text{ cP at } 35°C$$

Figure 6.4.1   Estimation of η_L (35°C) for 4-methyl-2-pentanone.

series of empirical coefficients. The latter have been calculated and listed, along with the applicable temperature range, for 314 compounds, including water, hydrocarbons, alcohols, ethers, aldehydes, ketones, acids, acetates, amines and substituted benzenes.

6.4.2   Compound-Independent Approaches: Totally Predictive Methods

Method of Joback and Reid   The method of Joback and Reid, discussed in Section 6.3, allows temperature-dependent estimation of viscosity based solely on molecular structure input.

Method of Mehrotra   The Mehrotra method [24] has been derived with 273 heavy ($M > 100$ g mol^{-1}) hydrocarbons such as n-paraffins, 1-olefins, branched paraffins and olefins, mono- and polycycloalkanes, and fused and nonfused aromatics. Based on 1300 individual dynamic viscosity–temperature values for these compounds, the following one-parameter equation has been obtained by employing regression analysis:

$$\log_{10}[\eta(\text{mPa·s}) + 0.8] = 100[0.01T]^b \qquad (6.4.5)$$

where T is in the range 283 to 473 K and b is a compound-specific parameter being tabulated [24]. The average absolute deviation (AAD) of eq. 6.4.5 for most compounds is under 10%, which has been reported to be well within the accepted precision for viscosity measurements. The parameter b can be calculated from the molar mass M using the following relationship:

$$b = Bm_0 + Bm_1 \cdot \log_{10} M \qquad (6.4.6a)$$

where Bm_0 and Bm_1, compound-class-specific coefficients, are given in Table 6.4.1. Equation 6.4.5 in combination with eq. 6.4.6a allows solely structure-based estimation of η. However, isomeric hydrocarbons even within each class have significantly different η values. These differences are not accounted for using the descriptor M. Therefore, correlations of b with the reduced boiling point have been derived:

$$b = Bt_0 + Bt_1 \cdot \log_{10} T_b^{10} \tag{6.4.6b}$$

where T_b^{10} is the boiling point in K at 10 mmHg. The coefficients Bt_0 and Bt_1 are listed in Table 6.4.2.

Grain's Method   Grain's method [1] requires the input of boiling point data, T_b and ΔH_{vb}. Grain proposes the estimation of ΔH_{vb} from T_b and structure-related K_F values (Fishtine), reducing the overall input to T_b, solely. In contrast, the model integrated in the Toolkit uses experimental ΔH_{vb} and T_b data from the database, applying the equation:

$$\ln \eta_L = \ln \eta_{Lb} + n^{-1}(\Delta H_{vb} - RT_x)(T_x^{-1} - T_b^{-1}) \tag{6.4.7}$$

where η_L and η_{Lb} are in cP, T_x and T_b in K, ΔH_{vb} is in cal mol^{-1}, R is 1.98723 cal (K·mol)$^{-1}$, and n is 8 for aliphatic hydrocarbons, 7 for ketones, and 5 for all other organic compounds. η_{Lb} is the viscosity at the boiling point, which is 0.4 for cyclohexane, 0.3 for benzene, 0.45 for alcohols and primary amines, and 0.2 for all

TABLE 6.4.1   Coefficients Bm_0 and Bm_1 and Correlation Coefficient r in Relationship 6.4.6a for Various Hydrocarbon Classes

Compound Class	Bm_0	Bm_1	r
n-Paraffins, 1-olefins	− 12.067	3.110	0.98
Branched paraffins and olefins	− 10.976	2.668	0.96
Nonfused aromatics	− 9.692	2.261	0.87
Fused-ring aromatics	− 9.309	2.185	0.82
Nonfused naphthenes	− 9.001	2.350	0.90
Fused-ring naphthenes	− 9.513	2.248	0.87

Source: Reprinted with permission from Ref. 24. Copyright (1991) American Chemical Society.

TABLE 6.4.2   Coefficients Bt_0 and Bt_1 and Correlation Coefficient r in Relationship 6.4.6b for Various Hydrocarbon Classes

Compound Class	Bt_0	Bt_1	r
n-Paraffins, 1-olefins	− 1.391	− 1.381	0.99
Branched paraffins and olefins	− 1.559	− 1.298	0.99
Nonfused aromatics	− 1.656	− 1.187	0.94
Fused-ring aromatics	− 1.722	− 1.099	0.86
Nonfused naphthenes	− 1.683	− 1.155	0.90
Fused-ring naphthenes	− 1.994	− 0.947	0.83

Source: Reprinted with pemission from Ref. 24. Copyright (1991) American Chemical Soiciety.

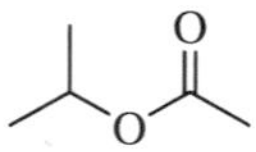

Isopropyl acetate

1. Classification: "other compound" $\rightarrow n = 5$
   "other compound" $\rightarrow \eta_{Lb} = 0.20$ cP

2. Boiling point: $T_b = 362.8$ K [26]

3. Enth. evaporation: $\Delta H_{vb} = 32.93$ kJ mol^{-1} [26]
   $\rightarrow \Delta H_{vb} = 7.865 \times 10^3$ cal mol^{-1}

4. With eq. 4.4.5: $\ln \eta = \ln 0.2 + 0.2[7865 - 1.98723(308.15)](308.15^{-1} - 362.8^{-1})$
   $= -0.9004$
   $\eta = 0.406$ cP at 35°C

Figure 6.4.2 Estimation of η at 35°C for isopropyl acetate. Data from Majer et al. [26].

other organic liquids. Values for ΔH_{vb} in J mol^{-1} have to be converted into cal mol^{-1} by multiplying by 0.238846. Grain's method is demonstrated in Figure 6.4.2 by estimating the viscosity of isopropyl acetate at 35°C. The corresponding experimental value is 0.4342 cP [25]. Grain's method is included in the Toolkit.

REFERENCES

1. Grain, C. F., Liquid Viscosity., in *Handbook of Chemical Property Estimation*, W. J. Lyman, W. F. Reehl, and D. H. Rosenblatt, Editors, 1990. Washington, DC: American Chemical Society.

2. Horvath, A. L., *Molecular Design: Chemical Structure Generation from the Properties of Pure Organic Compounds*, 1992. Amsterdam: Elsevier.

3. Rehberg, C. E., M. B. Dixon, and C. H. Fisher, Preparation and Physical Properties of *n*-Alkyl β-*n*-Alkoxypropionates. *J. Am. Chem. Soc.*, 1947: **69**, 2966–2970.

4. Rehberg, C. E. and M. B. Dixon, Mixed Esters of Lactic and Carbonic Acids: *n*-Alkyl Carbonates of Methyl and Butyl Lactates, and Butyl Carbonates of *n*-Alkyl Lactates. *J. Org. Chem.*, 1950: **15**, 565–571.

5. Rehberg, C. E., and M. B. Dixon, *n*-Alkyl Lactates and Their Acetates. *J. Am. Chem. Soc.*, 1950: **72**, 1918–1922.

6. Pachaiyappan, V., S. H. Ibrahim, and N. R. Kuloor, Simple Correlation for Determining Viscosity of Organic Liquids. *Chem. Eng.*, 1967: May 22, **66**, 193–196.

7. Lagemann, R. T., A Relation Between Viscosity and Refractive Index. *J. Am. Chem. Soc.*, 1945: **67**, 498–499.

8. Dixon, M. B., C. E. Rehberg, and C. H. Fisher, Preparation and Physical Properties of *n*-Alkyl-β-ethoxypropionates. *J. Am. Chem. Soc.*, 1948: **70**, 3733–3738.

9. Skubla, P., Prediction of Viscosity of Organic Liquids. *Collect. Czech. Chem. Commun.*, 1985: **50**, 1907–1916.

10. Cooper, E. F. and A.-F. A. Asfour, Densities and Kinematic Viscosities of Some C6–C16 *n*-Alkane Binary Liquid Systems at 293.15 K. *J. Chem. Eng. Data*, 1991: **36**, 285–288.

11. Haines, W. E., et al., Purification and Properties of Ten Organic Sulfur Compounds. *J. Phys. Chem.*, 1954: **58**, 270–278.

12. Morris, J. C., et al., Purification and Properties of Ten Organic Sulfur Compounds. *J. Chem. Eng. Data*, 1960: **5**, 112–116.

13. Tatevskii, V. M., et al., Relationship Between the Viscosity of Liquid and the Chemical Structure of Their Molecules. *Russ. J. Phys. Chem.*, 1985: **59**, 1643–1644.

14. Sounders, M., Jr., Viscosity and Chemical Constitution. *J. Am. Chem. Soc.*, 1938: **60**, 154–158.

15. Joback, K. G. and R. C. Reid, Estimation of Pure-Component Properties from Group-Contribution. *Chem. Eng. Commun.*, 1987: **57**, 233–243.

16. Riggio, R., et al., Viscosities, Densities, and Refractive Indexes of Mixtures of Methyl Isobutyl Ketone–Isobutyl Alcohol. *J. Chem. Eng. Data*, 1984: **29**, 11–13.

17. Reid, R. C., J. M. Prausnitz, and B. E. Poling, *The Properties of Gases and Liquids*, 4th ed., 1988, New York: McGraw-Hill.

18. Lide, D. R., and H. P. R. Frederikse, *CRC Handbook of Chemistry and Physics*, 75th ed., (1994–1995), 1994. Boca Raton, FL: CRC Press.

19. Riddick, J. A., *Organic Solvents: Physical Properties and Methods of Purification*, 4th ed., 1986. New York: Wiley.

20. Riggio, R., H. E. Martinez, and H. N. Solimo, Densities, Viscosities, and Refractive Indexes for the Methyl Isobutyl Ketone + Pentanols Systems: Measurements and Correlations. *J. Chem. Eng. Data*, 1986: **31**, 235–238.

21. Bingham, E. C., *Fluidity and Plasticity*, 1922. New York: McGraw-Hill.

22. Yaws, C. L., X. Lin, and L. Bu, Calculate Viscosities for 355 Liquids: Use the Temperature as a Starting Point. *Chem. Eng.*, 1994: Apr., 119–128.

23. Cao, W., et al., Group-Contribution Viscosity Predictions of Liquid Mixtures Using UNIFAC–VLE Parameters. *Ind. Eng. Chem. Res.*, 1993: **32**, 2088–2092.

24. Mehrotra, A. K., A Generalized Viscosity Equation for Pure Heavy Hydrocarbons. *Ind. Eng. Chem. Res.*, 1991: **30**, 420–427.

25. Krishnaiah, A., and D. S. Viswanath, Densities, Viscosities, and Excess Volumes of Isopropyl Acetate + Cyclohexane Mixtures at 298.15 and 308.15 K. *J. Chem. Eng. Data*, 1991: **36**, 317–318.

26. Majer, V., V. Svoboda, and H. V. Kehiaian, *Enthalpies of Vaporization of Organic Compounds. A Critical Review and Data Compilation*, Vol. 32, 1985. Oxford: Blackwell Scientific.

CHAPTER 7

VAPOR PRESSURE

7.1 DEFINITIONS AND APPLICATIONS

The vapor pressure, p_v, is the pressure exerted by fluids and solids at equilibrium with their own vapor phase. The vapor pressure is a strong function of T, as expressed in the Clausius–Clapeyron equation [1]:

$$\frac{dp_v}{dT} = \Delta H_v \left(\frac{RT^2}{p_{\text{vap}}}\right)^{-1} \Delta Z_v^{-1} \tag{7.1.1a}$$

or in the following form:

$$\frac{d\ln p_v}{d(1/T)} = \Delta H_v (R\Delta Z_v)^{-1} \tag{7.1.1b}$$

where ΔH_v is the enthalpy of vaporization, R the universal gas constant, and ΔZ_v a compressibility factor. Most vapor pressure–temperature correlations are derived by integrating eq. 7.1.1. The temperature dependence of the vapor pressure is discussed in further detail in Section 7.4.

CONVERSION OF PRESSURE UNITS

1 atm = 101.325 kPa = 760 torr = 760 mmHg
1 bar = 0.980665 atm
1 psia = 14.504 bar

The normal boiling point is defined as the temperature where the vapor pressure is 1 atm (760 mmHg). Under environmental conditions, the vapor pressures of liquid and solid compounds fall in the range 0 to 1 atm. Near-zero pressures are observed for high-boiling compounds with large molecular size and / or a high degree of molecular self-association. For example, DDT has a vapor pressure of 2×10^{-7} mmHg (at 20°C) and glycerol a vapor pressure of 3×10^{-3} mmHg (at 50°C) [2]. In contrast, the vapor pressure of n-hexane is 120 mmHg (at 20°C) and of benzene 76 mmHg (at 20°C) [2]. The vapor pressure of water at 25°C is 23.756 mmHg [3].

USES FOR VAPOR PRESSURE DATA

- To estimate liquid viscosity using Porter equation (6.2.1)
- To estimate the enthalpy of vaporization (Chapter 8)
- To estimate air–water partition coefficients (Chapter 12)
- To estimate rate of evaporation
- To estimate flash points using Affen's method [4]

7.2 PROPERTY–VAPOR PRESSURE RELATIONSHIPS

Method of Mackay, Bobra, Chan, and Shiu Mackay et al. [5] evaluated data of 72 solid and liquid halogenated and nonhalogenated hydrocarbons, all with boiling points above 100°C. Using data for 72 compounds, they derived the following equation from thermodynamic principles:

$$\ln\left[p_v(\text{atm})\right] = -(4.4 + \ln T_b)\left[1.803\left(\frac{T_b}{T} - 1\right) - 0.803 \ln\left(\frac{T_b}{T}\right)\right] - 6.8\left(\frac{T_m}{T} - 1\right) \tag{7.2.1}$$

where T, T_m, and T_b are in K. The third term including T_m is ignored for liquids, that is, when $T_m < T$.

Method of Mishra and Yalkowsky Mishra and Yalkowsky [6] have discussed the application of the method of Mackay (eq. 7.2.1). Based on the data set used by Mackay et al. [5] they derived the following model:

$$\ln\left[p_v(\text{atm})\right] = -\frac{T_m - T}{T}(8.5 - 5.0 \log \sigma_{\text{sym}} + 2.3 \log \phi_{\text{flx}})$$
$$- \frac{T_b - T}{T}(10 + 0.08 \log \phi_{\text{flx}})$$
$$+ \left(\frac{T_b - T}{T} - \ln \frac{T_b}{T}\right)(-6 - 0.9 \log \phi_{\text{flx}}) \tag{7.2.2}$$

where T, T_m, and T_b are in K, σ_{sym} is the rotational symmetry number, and ϕ_{flx} is the conformational flexibility number of the molecules. The structural parameter σ_{sym} is equal to the number of ways in which the molecule can be brought in positions that are identical with a reference position. For example, σ is 1 for ethylbenzene; 2 for toluene, o-xylene, m-xylene, 1,2,3-trimethylbenzene, and 1,2,4-trimethylbenzene; 4 for p-xylene and 1,2,4,5-tetramethylbenzene; and 6 for 1,3,5-trimethylbenzene. The structural parameter ϕ_{flx} is the number of reasonable conformations in which the molecule can exist. For rigid molecules, ϕ_{flx} is equal to 1. For linear alkanes and halogenated derivatives thereof, ϕ_{flx} is equal to 3^{N_c-3}. Examples are given by Mishra and Yalkowsky [6]. Application of eqs. 7.2.1 and 7.2.2 have been compared and the latter has been reported to give significantly better estimates using Mackay's data set as reference [6].

Solvatochromic Approach   Solvatochromic relationships are multivariate correlations between a property, usually solubility or partitioning property (see Sections 11.4 and 13.3), and *solvatochromic parameters*, parameters that account for the solutes interaction with the solvent. In the case of vapor pressure, the solvatochromic parameters only have to account for intermolecular interaction such as self-association between the solute (i.e., pure compound) molecules themselves. The following model has been reported for liquid and solid compounds, including hydrocarbons, halogenated hydrocarbons, alkanols, dialkyl ethers, and compounds such as dimethyl formamide, dimethylacetamide, pyridine, and dimethyl sulfoxide [7]:

$$\log_{10}[p_v(\text{mmHg})] = 7.82 - 7.29\left(\frac{V_I}{100}\right) - 6.41\pi^* + 3.25\pi^{*2} - 0.01(T_m - 25)$$

$$n = 53, \quad r = 0.98 \tag{7.2.3}$$

where p_{vap} is at 25°C, V_I is the intrinsic molar volume, π^* is a dipolarity–polarizability parameter and T_m is the melting point in °C. Liquids are assigned a T_m value of 25°C so that the melting point term in eq. 7.2.3 is diminished. Equation 7.2.3 has been derived with compounds in the $\log_{10}[p_{vap}(\text{mmHg})]$ range between -8 and $+4$.

Estimation of p_v for PCBs   Burkhard et al. [8] compared the predictive capability of 11 different methods to estimate p_{vap} for PCBs at 25°C. The comparison includes solely structure-based methods and methods that require the input of T_m, T_b, and the entropy of fusion, ΔS_{fus}, or a gas–liquid chromatographic retention index.

7.3   GROUP CONTRIBUTION APPROACHES FOR p_v

Method of Amidon and Anik   The method of Amidon and Anik [9] applicable to hydrocarbons and is based on the group additivity of the surface area. The approach is to model the Gibbs free energy change for the vaporization process, ΔG_v, as an additive parameter according to the following equation:

$$\Delta G_v = \Theta_0 + \sum \Theta_i(\text{GSA})_i \tag{7.3.1}$$

where Θ_0 is a constant, the Θ_i values are group-specific coefficients with units of $J \cdot mol^{-1} \cdot Å^{-2}$, $(GSA)_i$ is the group surface area contribution of the group of type i, and the summation is over all types i. GSA types are ArC (aromatic C atom), ArH (H attached to aromatic C atom), AlOV (aliphatic overlap), and ReAL (remaining aliphatic). The vapor pressure is related to ΔG_v as follows:

$$\ln p_v = -\frac{\Delta G_v}{RT} \qquad (7.3.2)$$

Burkhard et al. [8] have reported a correlation between ΔG_v and the surface area contributions of PCBs.

Method of Hishino, Zhu, Nagahama, and Hirata (HZNH)   The method of Hishino et al. [10] applies to mono-, di-, tri-, and tetra-substituted alkylbenzenes. Based on contributions for the groups $-CH_3$, $-CH_2-$, $>CH-$, and $>C<$ of the alkyl substituent and the aromatic ring groups $=CH-$ and $=C<$, the boiling points at three different pressures, p_1, p_2, and p_3 at 1.33 kPa, 101.32 kPa (=1 atm), and 199.98 kPa, respectively, are calculated. Using Thomson's method (see Section 7.4), the Antoine coefficients are derived, and with eq. 7.4.1 the vapor pressure at the desired temperature is estimated. The method error has been discussed, and application of this method has been demonstrated for 1,3-dimethyl-4-ethylbenzene and n-tetradecylbenzene [10].

Method of Macknick and Prausnitz   The method of Macknick and Prausnitz [11] allows estimation of vapor pressures for liquid hydrocarbons in the range 10 to 1500 mmHg. The method is based on the following equation proposed by Miller [12]:

$$\ln [p_v(\text{atm})] = A + \frac{B}{T} + C \ln T + DT + ET^2 \qquad (7.3.3)$$

where

$$A = \ln \frac{R}{V_W} + (s - 0.5) \ln \frac{E_0}{R} - \ln [(s-1)!] + \ln \alpha \qquad (7.3.3a)$$

$$B = -\frac{E_0}{R} \qquad (7.3.3b)$$

$$C = \frac{3}{2} - s \qquad (7.3.3c)$$

$$D = \frac{s-1}{E_0/R} \qquad (7.3.3d)$$

$$E = 0.5(s-3)(s-1)\left(\frac{E_0}{R}\right)^{-2} \qquad (7.3.3e)$$

where V_W is the van der Waals volume in $cm^3\,mol^{-1}$, E_0 the enthalpy of vaporization of the hypothetical liquid at 0 K, s the number of equivalent oscillators per molecule, R the gas constant ($82.06\ cm^3\ atm\ mol^{-1}\ K^{-1}$), α is equal to 0.0966, and E_0/R and T are in K. The parameters s, E_0/R, and V_W are often referred to as the AMP parameters, after the Abrams–Massaldi–Prausnitz model relating vapor liquid equilibria to kinetic theory. They can be calculated from group contributions using the scheme of Macknick and Prausnitz for s and E_0/R and the scheme of Bondi for V_W [11]. Edwards and Prausnitz [13] extended this method to include groups containing nitrogen and sulfur allowing estimations of vapor pressure in the range from 10 to 2000 mmHg. They demonstrate the extended method for carbazol and 2,3-dimethylthiophene. Burkhard [14] extended this method to include contribution for aromatic-ring-substituted halogens, F, Cl, Br, and I.

Method of Kelly, Mathias, and Schweighardt   The method of Kelly et al. [15] applies exclusively to perfluorinated saturated hydrocarbons over a pressure range of approximately 10 to 1000 mmHg. The method follows the approach of Macknick and Prausnitz. Group contributions for the three AMP parameters have been derived for the groups $-CF_3$, $-CF_2-$, $>CF-$ and $>C<$, and for extra contributions for five- and six-membered rings and ring fusion. These contributions apply to perfluoro compounds only. The model does not distinguish between differently substituted isomers. Application of this model has been demonstrated for perfluoroisopropylde-calin and perfluoroperhydrophenanthrene [15].

UNIFAC Approach   Jensen et al. [16] have employed the UNIFAC group contribution approach to develop an estimation method for pure-component vapor pressures. The model developed applies to hydrocarbons, alcohols, ketones, acids, and chloroalkanes of less than 500 molecular mass and in the vapor pressure region between 10 and 2000 mmHg. Burkhard et al. [8] extended this model to chlorinated aromatic compounds such as chlorobenzenes and PCBs.

7.4   TEMPERATURE DEPENDENCE OF p_v

The vapor pressure of a liquid increases with increasing temperature. Reviews on and discussion of different types of vapor pressure–temperature functions can be found in the literature [17–20]. The most common representation of vapor pressure–temperature data for a pressure interval of about 10 to 1500 mmHg [1] is the three-parametric Antoine equation:

$$\log p_v = A - \frac{B}{T + C} \qquad (7.4.1)$$

where the Antoine constants A, B, and C are compound-specific parameters. *A note of caution*: Units for vapor pressures and temperatures must correspond to those applicable for the Antoine constants. In Figures 7.4.1 and 7.4.2 the use of eq. 7.4.1 is demonstrated for estimation of the values p_v of tetrachloroethene at 46.5°C and of 2-methylpropanal at 36.1°C. An experimental value of 53.26 mmHg for tetra-chloroethene [22] and of 263.8 mmHg for 2-methylpropanal [23] have been reported at the specified temperatures.

Thomson's Method to Calculate Antoine Constants   The Thomson method utilizes available boiling temperatures at three different pressures to calculate

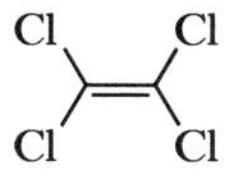

Tetrachloroethene

1. Antoine constant: A = 6.97683, B = 1386.92, C = 217.53 [24]

   Range: 37–120 °C

   Units: p_v in mmHg, T in °C

2. With eq. 6.4.1: $\log_{10} p_v = 6.97683 - \dfrac{1386.92}{46.5} + 217.53$

   $= 1.724$

   $p_v = 52.96$ mmHg at 46.5 °C

Figure 7.4.1 Estimation of p_v at 46.5°C for tetrachloroethene.

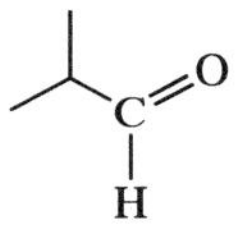

2-Methylpropanal

1. Antoine constant: A = 6.7351, B = 1053.2, C = 209.1 [24]

   Range : 13–63 °C

   Units : p_v in mmHg, T in °C

2. With eq. 6.4.1: $\log_{10} p_v = 6.7351 - \dfrac{1053.2}{36.1 + 209.1}$

   $= 2.440$

   $p_v = 275.32$ mmHg at 36.1 °C

Figure 7.4.2 Estimation of p_v at 36.1°C for 2-methylpropanal.

the Antoine coefficients. If the three boiling temperatures T_1, T_2, and T_3 are known at the pressures p_1, p_2, and p_3, respectively, the coefficient C is derived as follows [10]:

$$C = \frac{(T_3 - T_1) - T_3(1 - H)}{1 - H} \tag{7.4.2a}$$

where H is given as:

$$H = \frac{(\log p_3 - \log p_2)(T_2 - T_1)}{(\log p_2 - \log p_1)(T_3 - T_2)}$$

Then B and A are calculated as follows:

$$B = \frac{(\log p_3 - \log p_1)(T_1 + C)(T_3 + C)}{T_3 - T_1} \qquad (7.4.2b)$$

$$A = \log p_1 + \frac{B}{C + T_1} \qquad (7.4.2c)$$

Methods Based on the Frost–Kalkwarf Equation   The Frost–Kalkwarf equation relies on four compound-specific coefficients to correlate vapor pressure–temperature data:

$$\log p_v = A + \frac{B}{T} + C \log T + D p_v T^{-2} \qquad (7.4.3)$$

The four coefficients A, B, C, and D have been derived, for example, with selected hydrocarbons [25, 26]. Equation 7.4.3 accurately represents the vapor pressure function over the entire temperature range between the triple point and the critical point. If the coefficients are not available for a given compound, they can be calculated. D is calculated from the pressure van der Waals constant, a, which can be estimated from group contributions. B is calculated directly from group contributions. Then the coefficients A and C can be estimated from two p_v/T points (e.g., normal boiling point and critical point). This approach has been evaluated for various classes of hydrocarbons commonly encountered in petroleum technology [25, 26].

Methods to Estimate p_v from T_b Only   The modified Watson correlation [2] applies for liquids and solids in the p_v range from 10^{-7} to 760 mmHg. This method is based on the Watson equation (8.5.1) and requires the input of the normal boiling point temperature, T_b, and of ΔH_{vb}. However, the latter property is itself calculated from T_b and from structural parameters. For compounds with p_v between 10 and 760 mmHg, a method error of 2.5% has been reported, whereas a considerably higher error has been found for compounds with p_v below 10 mmHg. The method has been illustrated for benzene and DDT [2]. A large number of other T_b/p_v correlations have been discussed by Horvath [17].

Methods to Estimate p_v Solely from Molecular Structure   Methods of this type are available with the GCM approaches. All methods presented in Section 7.3 allow temperature-dependent estimation of p_v in the region specified. For certain homologous series, specific vapor pressure–structure–temperature relationships exist. For example, Woodman et al. [27] have reported the following relationship for α, ω-dinitriles ($3 < N_{CH_2} < 8$):

$$\log_{10} p_v = 11.936 + 0.222 N_{CH_2} - \frac{173 N_{CH_2} + 2996}{T} \qquad (7.4.4)$$

where p_v is in microns and T is in K. Most of the vapor pressures calculated from Eq. 7.4.4 agreed with the experimental values to within 2%, and all of the calculated values agreed within 4%.

REFERENCES

1. Reid, R. C., J. M. Prausnitz, and T. K. Sherwood, *The Properties of Gases and Liquids*, 3rd ed., 1977. New York: McGraw-Hill.

2. Grain, C. F., Vapor Pressure, in *Handbook of Chemical Property Estimation*, W. J. Lyman, W. F. Reehl, and D. H. Rosenblatt, Editors, 1990. Washington, DC: American Chemical Society.

3. Lide, D. R., and H. P. R. Frederikse, *CRC Handbook of Chemistry and Physics*, 75th ed. (1994–1995), 1994. Boca Raton, FL: CRC Press.

4. Hagopian, J. H., Flash Points of Pure Substances, in *Handbook of Chemical Property Estimation*, W. J. Lyman, W. F. Reehl, and D. H. Rosenblatt, Editors, 1990. Washington, DC: American Chemical Society.

5. Mackay, D., A. Bobra, D. W. Chan, and W. Y. Shiu, Vapor Pressure Correlation for Low-Volatility Environmental Chemicals. *Environ. Sci. Technol.*, 1982: **16**, 645–649.

6. Mishra, D. S., and S. H. Yalkowsky, Estimation of Vapor Pressure of Some Organic Compounds. *Ind. Eng. Chem. Res.*, 1991: **30**, 1609–1612.

7. Banerjee, S., P. H. Howard, and S. S. Lande, General Structure–Vapor Pressure Relationships for Organics. *Chemosphere*, 1990: **21**, 1173–1180.

8. Burkhard, L. P., A. W. Andren, and D. E. Armstrong, Estimation of Vapor Pressures for Polychlorinated Biphenyls: A Comparison of Eleven Predictive Methods. *Chemosphere*, 1985: **19**, 500–507.

9. Amidon, G. L., and S. T. Anik, Application of the Surface Area Approach to the Correlation and Estimation of Aqueous Solubility and Vapor Pressure: Alkyl Aromatic Hydrocarbons. *J. Chem. Eng. Data*, 1981: **26**, 28–33.

10. Hishino, D., et al., Prediction of Vapor Pressures for Substituted Benzenes by a Group-Contribution Method. *Ind. Eng. Chem. Fundam.*, 1985: **24**, 112–114.

11. Macknick, A. B., and J. M. Prausnitz, Vapor Pressures of Heavy Liquid Hydrocarbons by a Group Contribution Method. *Ind. Eng. Chem. Fundam.*, 1979: **18**, 348–351.

12. Miller, D. G., Derivation of Two Equations for the Estimation of Vapor Pressures. *J. Phys. Chem.*, 1964: **68**, 1399–1408.

13. Edwards, D. R., and J. M. Prausnitz. Estimation of Vapor Pressures of Heavy Liquid Hydrocarbons Containing Nitrogen or Sulfur by a Group-Contribution Method. *Ind. Eng. Chem. Res.*, 1981: **20**, 280–283.

14. Burkhard, L. P., Estimation of Vapor Pressures for Halogenated Aromatic Hydrocarbons by a Group-Contribution Method. *Ind. Eng. Chem. Fundam.*, 1985: **24**, 119–120.

15. Kelly, C. M., P. M. Mathias, and F. K. Schweighardt, Correlating Vapor Pressures of Perfluorinated Saturated Hydrocarbons by a Group Contribution Method. *Ind. Eng. Chem. Res*, 1988: **27**, 1732–1736.

16. Jensen, T., A. Fredenslund, and P. Rasmussen, Pure-Component Vapor Pressures Using UNIFAC Group Contribution. *Ind. Eng. Chem. Fundam.*, 1981: **20**, 239–246.

17. Horvath, A. L., *Molecular Design: Chemical Structure Generation from the Properties of Pure Organic Compounds*, 1992. Amsterdam: Elsevier.

18. Somayajulu, G. R., New Vapor Pressure Equations from Triple Point to Critical Point and a Predictive Procedure for Vapor Pressure. *J. Chem. Eng. Data*, 1986: **31**, 438–447.

19. Riddick, J. A., *Organic Solvents: Physical Properties and Methods of Purification*, 4th ed., 1986. New York: Wiley.

20. Ambrose, D., J. F. Counsell, and C. P. I. Hicks, A new procedure for estimation and extrapolation. *J. Chem. Thermodynamics*, 1978: **10**, 771–778.

21. Ambrose, D., The Correlation and estimation of vapor pressures: I. A. comparison of the three vapor-pressure equations. *J. Chem. Thermodynamics*, 1978: **10**, 765–769.

22. Polak, J., and B. C.-Y. Lu, Mutual Solubilities of Hydrocarbons and Water at 0 and 25°C. *Can. J. Chem.*, 1973: **51**, 4018–4023.

23. Brazhnikov, M. M., A. D. Peshchenko, and O. V. Ralko, Heats of Vaporization of C1-C4 Aliphatic-Aldehydes and of Dimethoxymethane. *J. of Appl. Chem. of the USSR*, 1976: **49**(5), 1983–1085.

24. Dean, J. A., *Lange's Handbook of Chemistry*, 14th ed., 1992. New York: McGraw-Hill.

25. Bond, D. L., and G. Thodos, Vapor Pressures of Alkyl Aromatic Hydrocarbons. *J. Chem. Eng. Data*, 1960: **5**, 289–292.

26. Pasek, G. J., and G. Thodos, Vapor Pressures of the Naphthenic Hydrocarbons. *J. Chem. Eng. Data*, 1962: **7**: 21–26.

27. Woodman, A. L., W. J. Murbach, and M. H. Kaufman, Vapor Pressure and Viscosity Relationships for a Homologous Series of α, ω-Dinitriles. *J. Phys. Chem.*, 1960: **64**, 658–660.

CHAPTER 8

ENTHALPY OF VAPORIZATION

8.1 DEFINITIONS AND APPLICATIONS

The *enthalpy of vaporization*, ΔH_v, is defined as the difference between the vapor- and liquid-phase enthalpies at a given temperature and the corresponding saturated vapor pressure [1]:

$$\Delta H_v = (H^g - H^l)_{\text{sat}} \qquad (8.1.1)$$

where both phases are at equilibrium. ΔH_v is defined for liquids from the triple point to the critical point. It diminishes at the critical point.

CONVERSION OF ENTHALPY OF VAPORIZATION UNITS

$$1\,\text{cal} = 0.238846\,\text{J}$$
$$1\,\text{J} = 4.1868\,\text{cal}$$

USES FOR DATA ON ENTHALPY OF VAPORIZATION

- To estimate surface tension
- To estimate liquid viscosity [2]
- To estimate liquid viscosity with Grain's method (eq. 6.4.7)
- To derive cohesion parameters [3]

With ΔH_v given at temperature T, the entropy of vaporization, ΔS_v, is obtained as follows:

$$\Delta S_v = \frac{\Delta H_v}{T} \qquad (8.1.2)$$

8.2   PROPERTY–ΔH_v RELATIONSHIPS

T_b–ΔH_v Relationships   Trouton (1884) proposed that for liquids, ΔS_v is a constant [4]. This rule is commonly applied at the normal boiling point, T_b. *Trouton's rule* at normal boiling point is

$$\Delta S_{vb} = \frac{\Delta H_{vb}}{T_b} = \text{constant} \tag{8.2.1}$$

where ΔH_{vb} and ΔS_{vb} are the enthalpy and entropy of vaporization at T_b, respectively.

A series of T_b–ΔH_v correlations, based on the refinement of Trouton's rule, have been published. Horvath [5] reviews proposed T_b–ΔH_{vb} relationships and T_b–ΔH_v relationships where ΔH_v is at 25°C. When applying these relationships to given compounds, one has to consider possible vapor-phase intermolecular associations.

Critical Point–ΔH_v Relationships   Estimation methods that use critical point data to estimate ΔH_v have been reviewed in various accounts [1,5,6]. Usually, additional input such as boiling point data T_b and ΔH_b or the acentric factor is required. The advantage of these methods is that they allow temperature-dependent estimation. One such method is presented in further detail in Section 8.5.

8.3   STRUCTURE–ΔH_v RELATIONSHIPS

Most of the structure–ΔH_v relationships have been developed for either ΔH_v at 25°C or for the normal boiling point enthalpy, ΔH_{vb}. Relationships of both types are discussed below.

Homologous Series   Månsson et al. [7] have reported linear relationships between ΔH_v at 25°C and the number of methylene groups:

$$\Delta H_v[\text{X}-(\text{CH}_2)_m-\text{H}](\text{kJ mol}^{-1}) = A + Bm \tag{8.3.1}$$

The coefficients A and B along with their standard deviations s_A and s_B are shown in Table 8.3.1. Equation 8.3.1 is invalid for nitrilo- and acetylalkanes. Homologous series with multiple functional groups, such as dialkyl sebacate esters and triglyceride esters, have been studied by Kishore et al. [8].

Woodman et al. [9] have derived the following relationship for α,ω-dinitriles ($2 < N_{\text{CH}_2} < 8$):

$$\Delta H_v = (0.79 \pm 0.02)N_{\text{CH}_2} + (13.71 + 0.11) \tag{8.3.2}$$

where ΔH_v is in kcal mol^{-1} and in the temperature range 5 to 65°C.

Piacente et al. [10] observed an odd–even effect for ΔH_v at 25°C considering high-molar-mass *n*-alkanes with $N_C > 19$. They obtained the following equations for even *n*-alkanes ($19 < N_C < 39$):

$$\Delta H_v(\text{kJ mol}^{-1}) = (63 \pm 4) + (2.62 \pm 0.14)N_C \tag{8.3.3a}$$

TABLE 8.3.1   Coefficients A and B of Eq. 8.3.1 with Statistical Parameters for Various Homologous Series [7]

X	n^a	$A \pm s_A$	$B \pm s_B$	s_0^b	Valid for $m \geq$
Hydrogen	18	1.89 ± 0.07	4.953 ± 0.006	0.091	5
Vinyl	5	10.73 ± 0.30	4.972 ± 0.033	0.276	3
Hydroxy	16	32.43 ± 0.19	4.937 ± 0.026	0.379	2
Mercapto	5	17.70 ± 0.28	4.760 ± 0.050	0.310	2
Chloro	8	13.85 ± 0.18	4.854 ± 0.021	0.242	3
Bromo	8	17.36 ± 0.13	4.803 ± 0.015	0.170	3
Methoxycarbonyl	12	21.93 ± 0.64	5.029 ± 0.069	0.747	4

$^a n$ is the number of members from the homologous series.
$^b s_0 = \{\sum[\Delta H_v(\text{obs.}) - \Delta H_v(\text{calc.})]^2/(n-2)\}^{-1/2}$.

Source: Ref. 7. Reprinted with permission, copyright (1977) Academic Press.

and for odd n-alkanes ($20 < N_C < 38$):

$$\Delta H_v(\text{kJ mol}^{-1}) = (57 \pm 6) + (2.74 \pm 0.22)N_C \tag{8.3.3b}$$

Chain-Length Method of Mishra and Yalkowsky   The method of Mishra and Yalkowsky [11] is a modification of Trouton's rule for long-chain hydrocarbons, including alkanes, alkenes, cyclopentanes, cyclohexanes, and alkylbenzenes. The relationship is

$$\frac{\Delta H_{vb}}{T_b} = 20 + 0.16(N_{CH_2,\text{chain}} - 5) \tag{8.3.4}$$

for $N_{CH_2,\text{chain}} \leq 5$, where $N_{CH_2,\text{chain}}$ is the number of $-CH_2-$ groups in the chain. $\Delta H_{vb}/T_b$ is in cal K^{-1} mol^{-1}.

Geometric Volume–ΔH_v Relationship   Bhattacharjee and Dasgupta [12] studied correlations between ΔH_v and the geometric volume. They reported the following bilinear relationship for alkanes ($C_1–C_8$):

$$\Delta H_v(\text{cal g}^{-1}) = N_C(16.1869V_g - 39.113) + (-148.917V_g + 574.0554) \tag{8.3.5}$$

where N_C is the number of carbon atoms per molecule and V_g is the geometric volume.

Wiener–Index–ΔH_{vb} Relationship   Bonchev et al. [13] have reported the following relationship for alkanes ($C_2–C_{10}$):

$$\Delta H_{vb}(\text{kJ mol}^{-1}) = 19.0221 + 13.6898N_C - 0.9397(N_C)^2 + 0.0101W$$
$$n = 15, \quad s = 1.53, \quad r = 0.9946 \tag{8.3.6}$$

Molecular Connectivity–ΔH_{vb} Relationship Kier and Hall [14] derived the following relationship for alkanes (C_2–C_{16}):

$$\Delta H_{vb}(\text{kcal mol}^{-1}) = 0.567 + 1.752\,{}^{1}\chi + 1.301\,{}^{2}\chi - 0.415\,{}^{4}\chi_p - 0.437\,{}^{5}\chi_p$$
$$- 0.266\,{}^{6}\chi_p - 2.091\,{}^{3}\chi_c + 3.051\,{}^{4}\chi_c$$
$$n = 44, \quad s = 0.042, \quad r = 0.9999 \tag{8.3.7}$$

Similar relationships have been reported by the same authors for alcohols.

Needham et al. [15] derived the following model for alkanes (C_2–C_9):

$$\Delta H_v(\text{kJ mol}^{-1}) = 1.9(\pm 0.3) + 4.5(\pm 0.2)\,{}^{0}\chi + 4.8(\pm 0.4)\,{}^{2}\chi - 12.4(\pm 0.5)\,{}^{3}\chi_c$$
$$+ 20.7(\pm 0.8)\,{}^{4}\chi_c + 0.25(\pm 0.05)\,{}^{4}\chi_{pc}$$
$$n = 69, \quad s = 0.2, \quad r^2 = 0.998, \quad F = 7849 \tag{8.3.8}$$

where ΔH_v is at 25°C.

White [16] has derived an univariate relationship for PAHs:

$$\Delta H_{vb}(\text{kJ mol}^{-1}) = 25.147 + 6.464\,{}^{1}\chi_v \qquad n = 47, \quad s = 9.39, \quad r = 0.993 \tag{8.3.9}$$

Molar Mass–ΔH_{vb} Relationship Ibrahim and Kuloor [17] proposed the following equation:

$$\Delta H_{vb}(\text{cal g}^{-1}) = CM^{-n} \tag{8.3.10}$$

where C and n are empirical, compound-class-specific constants given in Table 8.3.2. This model is based on 160 compounds with M values ranging from 16 to 240

TABLE 8.3.2 Coefficients C and n in Relationship 8.3.10 for Various Hydrocarbon Classes

Compound Class	C	n
Aliphatic hydrocarbons	367	0.342
Cyclic hydrocarbons	605	0.440
Aromatics	1155	0.574
Halogenated aliphatics	1280	0.650
Alcohols	3475	0.745
Ethers	315	0.300
Aldehydes, oxides, anhydrides	940	0.494
Ketones	3200	0.795
Acids	7200	0.930
Esters	1550	0.642
Aliphatic amines	3250	0.825

Source: Ref. 17. Reprinted with permisson. Copyright (1966) Chemical Engineering.

$g\,mol^{-1}$. The overall error has been reported as 2%; approximately 100 compounds fit accurately, and 40 compounds are within 3%.

8.4   GROUP CONTRIBUTION APPROACHES FOR ΔH_v

Various GCMs are available to estimate ΔH_v. A comprehensive discussion of several has been given by Horvath [5]. Five selected methods are presented here.

Method of Garbalena and Herndon   The Garbalena and Herndon model applies to alkanes (C_2–C_{15}) [18]. It is based on atom contribution and contributions of atom pairs in which the atoms are two bonds apart. The GCM equation is:

$$\Delta H_v(kJ\,mol^{-1}) = (5.83 \pm 0.07)N_{CH_3} + (4.94 \pm 0.04)N_{CH_2} + (2.69 \pm 0.17)N_{CH}$$
$$- (1.55 \pm 0.80)N(C, CH)_2 - (1.39 \pm 0.87)N(C, C)_2$$
$$n = 42, \quad s.e. = 0.772, \quad r = 0.997, \quad F = 1780 \qquad (8.4.1)$$

where C, CH, CH_2, and CH_3 represent a quaternary carbon, a tertiary carbon, a methylene group, and a methyl group, respectively. $N(C,CH)_2$ is the number of C−CH pairs and $N(C,C)_2$ is the number of C−C pairs. The subscript 2 indicates that the groups are two bonds apart. This model may be interpreted as a atom contribution model with two correction terms ($N(C,CH)_2$, $N(C,C)_2$) for multiple-branched molecules. Note that these two contribution have negative coefficients, indicating a decrease in ΔH_v between isomers with increasing "branchedness", which is consistent with the experimental data.

Method of Ma and Zhao   The Ma and Zhao method [19] can be applied to estimate the entropy of vaporization at the normal boiling point, ΔS_{vb}. The method has been developed from a set of 483 compounds, including alkanes, alkenes, alkynes, cyclic hydrocarbons, aromatic hydrocarbons and derivatives, halogenated hydrocarbons, alcohols, aldehydes, ketones, esters, ethers, and multioxygen-, oxygen–halogen-, nitrogen-, and sulfur-containing compounds. Groups are classified as nonring, in-ring, connected-to-ring, and aromatic-ring groups. A total of 94 group contributions, $(\Delta S_{vb})_i$, have been derived. The model equation is:

$$\Delta S_{vb} = 86.9178\,J\,mol^{-1}\,K^{-1} + \sum n_i(\Delta S_{vb})_i \qquad (8.4.2)$$

where ΔS_{vb} and $(\Delta S_{vb})_i$ are in $J\,mol^{-1}\,K^{-1}$ and n_i is the occurrence frequency of group i. Many of the contributions can be found between -2 and $+2\,J\,mol^{-1}\,K^{-1}$. For compounds that contain only groups of this kind, the summation term in eq. 8.4.2 will be approximately zero (i.e. ΔS_{vb} is approximately constant for these compounds, in concordance with Trouton's rule). In contrast, all hydroxyl group contributions exceed $15\,J\,mol^{-1}\,K^{-1}$, demonstrating a significant deviation from Trouton's rule for compounds containing these groups. The highest contribution is 21.97611 $J\,mol^{-1}\,K^{-1}$ for *t*-COH and the lowest contribution is -4.02309 for −$CFCl_2$. The average prediction error is 1.4%. The greatest average error, 3.1%, is found for alcohols and has been attributed to their strong hydrogen-bond effect. Extensive comparisons between this and methods derived previously, including GCMs and corresponding-state methods, has been made. Estimation examples have been given for 1-methyl-1-ethylcyclopentane and naphthalene.

Method of Hishino, Zhu, Nagahama, and Hirata (HZNH)   The method of Hishino et al. [20 and references cited therein] applies to mono-, di-, tri-, and tetrasubstituted alkylbenzenes (compare with the corresponding method in Section 7.3). It is derived from eqs. 7.1.1 and 7.3.3:

$$\Delta H_v(\text{kcal mol}^{-1}) = R(-B + CT + DT^2 + 2ET^3) \qquad (8.4.3)$$

where T is in K and the coefficients B, C, D, and E are calculated with eqs. 7.3.3b to e, respectively. For 67 liquids, an average error in ΔH_{vb} of $\pm 5.4\%$ has been reported.

Method of Joback and Reid   The method of Joback and Reid [21] applies to ΔH_v estimation at the normal boiling point only. It has been derived from a database of 368 compounds and yielded an average absolute error of 1.27 kJ mol^{-1} corresponding to a 3.9 average percent error using the training set ΔH_{vb} values. The GCM equation is:

$$\Delta H_{vb}(\text{kJ mol}^{-1}) = 15.30 + \sum n_i(\Delta H_{vb})_i \qquad n = 368, \quad s = 1.79 \,\text{kJ mol}^{-1}$$
$$(8.4.4)$$

where the summation is over all group types i. $(\Delta H_{vb})_i$ is the contribution for the ith group type and n_i is the number of times the group occurs in the molecule.

Method of Constantinou and Gani   The approach of Constantinou and Gani [22] has been described for T_b in Section 9.3. The analog model for ΔH_v at 25°C is:

$$\Delta H_v - 6.829 \,\text{kJ mol}^{-1} = \sum n_i(\Delta H_{v1})_i + W \sum m_j(\Delta H_{v2})_j \qquad n = 225$$
$$W = 0 : s = 2.20 \,\text{kJ mol}^{-1}, \quad \text{AAE} = 1.40 \,\text{kJ mol}^{-1}, \quad \text{AAPE} = 3.22\%$$
$$W = 1 : s = 1.83 \,\text{kJ mol}^{-1}, \quad \text{AAE} = 1.11 \,\text{kJ mol}^{-1}, \quad \text{AAPE} = 2.57\% \; (8.4.5)$$

where $(\Delta H_{v1})_i$ is the contribution of the first-order group type i, which occurs n_i times in the molecule, and $(\Delta H_{v2})_j$ is the contribution of the second-order type j with m_j occurrences in the molecule. W is zero or 1 for a first- and second-order approximation, respectively and the statistical parameters are $s = [\sum(T_{b,\text{fit}} - T_{b,\text{obs}})^2/n]^{1/2}$, $\text{AAE} = (1/n) \sum |T_{b,\text{fit}} - T_{b,\text{obs}}|$, and $\text{AAPE} = (1/n) \sum |T_{b,\text{fit}} - T_{b,\text{obs}}|/T_{b,\text{obs}} \times 100\%$.

8.5   TEMPERATURE DEPENDENCE OF ΔH_v

The enthalpy of vaporization decreases as the temperature increases. The only exceptions are compounds that undergo strong association in the vapor phase. The temperature dependence of ΔH_v has been reviewed by Majer [23] and Tekác [24].

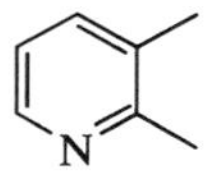

2,3-Dimethylpyridine

1. Input properties:
$\Delta H_{vb} = 39.08 \text{ kJ mol}^{-1}$
$T_b = 434.4 \text{ K}$
$T_c = 655.4 \text{ K}$

2. Fishtine n:
$\dfrac{T_b}{T_c} = \dfrac{434.4}{655.4} = 0.663$

$n = 0.74\,(0.663) - 0.116 = 0.374 \text{ (eq. 8.5.1b)}$

3. Using eq. 8.5.1:
$\Delta H_v = 39.08 \left(\dfrac{1 - 298.2/655.4}{1 - 0.663} \right)^{0.374}$

$= 39.08\,(1.616^{0.374})$
$= 46.8 \text{ kJ mol}^{-1} \text{ at } 25°C$

Figure 8.5.1 Estimation of ΔH_v at 25°C for 2,3-dimethylpyridine using input properties from Majer et al. [1].

This section is limited to the method of Watson [25]. Watson's method allows the estimation of ΔH_v at a given T, if T_b, T_c, and ΔH_{vb} are known. The Watson equation is

$$\Delta H_v = \Delta H_{vb} \left(\frac{1 - T/T_c}{1 - T_b/T_c} \right)^n \tag{8.5.1}$$

where T, T_c, and T_b are in K, ΔH_v and ΔH_{vb} are in cal mol^{-1} or J mol^{-1} and n is equal to 0.38. Fishtine proposed the following n values that yield better estimates of ΔH_v:

$$n = \begin{cases} 0.30 & \text{if } \dfrac{T_b}{T_c} < 0.57 & (8.5.1a) \\[2ex] 0.74\left(\dfrac{T_b}{T_c}\right) - 0.116 & \text{if } 0.57 \le \dfrac{T_b}{T_c} \le 0.71 & (8.5.1b) \\[2ex] 0.41 & \text{if } \dfrac{T_b}{T_c} > 0.71 & (8.5.1c) \end{cases}$$

Equation 8.5.1 has been implemented in the Toolkit using the n values of Fishtine. The program calculates ΔH_v based on ΔH_{vb}, T_c, and T_b data compiled by Majer et al. [1]. The method is illustrated for 2,3-dimethylpyridine at 25 K in Figure 8.5.1. An experimental value of 47.786 kJ mol^{-1} [26] has been reported.

REFERENCES

1. Majer, V., V. Svoboda, and H. V. Kehiaian, *Enthalpies of Vaporization of Organic Compounds: A Critical Review and Data Compilation*, Vol. 32, 1985. Oxford: Blackwell Scientific.

2. Cao, W., A. Fredenslund, and P. Rasmussen, Statistical Thermodynamic Model for Viscosity of Pure Liquids and Liquid Mixtures. *Ind. Eng. Chem. Res.*, 1992: **31**, 2603–2619.

3. Barton, A. F. M., *CRC Handbook of Solubility and Other Cohesion Parameters*, 2nd ed., 1991. Boca Raton, FL: CRC Press.

4. Shinoda, K., Entropy of Vaporization at the Boiling Point. *J. Chem. Phys.*, 1983: **78**, 4784.

5. Horvath, A. L., *Molecular Design: Chemical Structure Generation from the Properties of Pure Organic Compounds*, 1992. Amsterdam: Elsevier.

6. Reid, R. C., J. M. Prausnitz, and T. K. Sherwood, *The Properties of Gases and Liquids*, 3rd ed., 1977. New York: McGraw-Hill.

7. Månsson, M., et al., Enthalpies of Vaporization of Some 1-Substituted *n*-Alkanes. *J. Chem. Thermodyn.*, 1977: **9**, 91–97.

8. Kishore, K., H. K. Shobha, and G. J. Mattamal, Structural Effects on the Vaporization of High Molecular Weight Esters. *J. Phys. Chem.*, 1990: **94**, 1642–1648.

9. Woodman, A. L., W. J. Murbach, and M. H. Kaufman, Vapor Pressure and Viscosity Relationships for a Homologous Series of α,ω-Dinitriles. *J. Phys. Chem.*, 1960: **64**, 658–660.

10. Piacente, V., D. Fontana, and P. Scardala, Enthalpies of Vaporization of a Homologous Series of *n*-Alkanes Determined from Vapor Pressure Measurement. *J. Chem. Eng. Data*, 1994: **39**, 231–237.

11. Mishra, D. S., and S. H. Yalkowsky, Estimation of Entropy of Vaporization: Effect of Chain Length. *Chemosphere*, 1990: **21**, 111–117.

12. Bhattacharjee, S., and P. Dasgupta, Molecular Property Correlation in Alkanes with Geometric Volume. *Comput. Chem.*, 1994: **18**, 61–71.

13. Bonchev, D., V. Kamenska, and O. Mekenyan, Comparability Graphs and Molecular Properties: IV. Generalization and Application. *J. Math. Chem.*, 1990: **5**, 43–72.

14. Kier, L. B., and L. H. Hall, *Molecular Connectivity in Chemistry and Drug Research*, 1976. San Diego, CA: Academic Press.

15. Needham, D. E., I.-C. Wei, and P. G. Seybold, Molecular Modeling of the Physical Properties of the Alkanes. *J. Am. Chem. Soc.*, 1988: **110**, 4186–4194.

16. White, C. M., Prediction of the Boiling Point, Heat of Vaporization, and Vapor Pressure at Various Temperatures for Polycyclic Aromatic Hydrocarbons. *J. Chem. Eng. Data*, 1986: **31**, 198–203.

17. Ibrahim, S. H., and N. R. Kuloor, Use of Molecular Weight to Estimate Latent Heat. *Chem. Eng.*, 1966. Dec. 5; 147–148.

18. Garbalena, M., and W. C. Herndon, Optimum Graph-Theoretical Models for Enthalpic Properties of Alkanes. *J. Chem. Inf. Comput. Sci.*, 1992: **32**, 37–42.

19. Ma, P., and X. Zhao, Modified Group Contribution Method for Predicting the Entropy of Vaporization at the Normal Boiling Point. *Ind. Eng. Chem. Res.*, 1993: **32**, 3180–3183.

20. Hishino, D., et al., Prediction of Vapor Pressures for Substituted Benzenes by a Group-Contribution Method. *Ind. Eng. Chem. Fundam.*, 1985: **24**, 112–114.

21. Joback, K. G., and R. C. Reid, Estimation of Pure-Component Properties from Group-Contribution. *Chem. Eng. Commun.*, 1987: **57**, 233–243.

22. Constantinou, L., and R. Gani, New Group Contribution Method for Estimating Properties of Pure Compounds. *AIChE J.*, 1994: **40**, 1697–1710.

23. Majer, V., Enthalpy of Vaporization Basic Relations and Major Applications, in *Enthalpies of Vaporization of Organic Compounds. A Critical Review and Data Compilation*, V. Majer, V. Svoboda, and H. V. Kehiaian, Editors, 1985. Oxford: Blackwell Scientific.

24. Tekác, V., et al., Enthalpies of Vaporization and Cohesive Energies for Six Monochlorinated Alkanes. *J. Chem. Thermodyn.*, 1981: **13**, 659–662.

25. Rechsteiner, C. E., Heat of Vaporization, in *Handbook of Chemical Property Estimation*, W. J. Lyman, W. F. Reehl, and D. H. Rosenblatt, Editors, 1990. Washington, DC: American Chemical Society.

26. Wisniewska, B., M. Lencka, and M. Rogalski, Vapor Pressures of 2,4-, 2,6-, and 3,5-Dimethylpyridine at Temperatures from 267 to 360 K. *J. Chem. Thermodyn.*, 1986: **18**, 703–708.

CHAPTER 9

BOILING POINT

9.1 DEFINITIONS AND APPLICATIONS

The boiling point is the temperature at which the vapor pressure of a liquid equals the pressure of the atmosphere on the liquid [1]. The normal boiling point, T_b, is the boiling point at the pressure of 1 atm ($=101.325\,N\,m^{-2}$). Impurities in the liquid can change the boiling temperature. Reported experimental T_b values are usually below 300°C, because decomposition occurs for most compounds at higher temperatures, if not already below. Distillation of compounds with "virtually high T_b" is performed under reduced pressure.

CONVERSION OF TEMPERATURE UNITS

degree Celsius: $°C = K - 273.15$

Kelvin: $K = °C + 273.15$

Kelvin: $K = [(\frac{5}{9})(°F - 32)] + 273.15$

degree Fahrenheit: $°F = (\frac{9}{5})(K - 273.15) + 32$

USES FOR BOILING POINT DATA

- To indicate (together with the melting point) the physical state of a compound
- To measure the purity of a compound
- To assess the volatility of liquids
- To estimate liquid viscosity with Grain's method (eq. 6.4.7)
- To estimate vapor pressure using QPPRs (eqs. 7.2.1 and 7.2.2)

- To estimate vapor pressure with the modified Watson approach (Section 7.4)
- To estimate enthalpy of vaporization with Watson's equation 8.5.1
- To estimate aqueous solubility using QPPRs (eqs. 11.4.11 to 11.4.13)
- To estimate flash points: T_f(cc) and T_f(oc) [2,3]
- To model thermal conductivity of liquid mixtures [4]

Guldberg Ratio   The normal boiling point divided by the critical temperature is the *Guldberg ratio*:

$$\Theta = \frac{T_b}{T_c} \qquad (9.1.1)$$

where T_b and T_c have to be in K.

9.2 STRUCTURE–T_b RELATIONSHIPS

Relationships between T_b and N_C or M in homologous series are nonlinear. The difference in T_b between successive members of *n*-alkanes is not constant. It falls off continuously, demonstrated by plotting T_b against N_C (C_1–C_{40}) [5]. The following equation has been reported for *n*-alkanes (C_6–C_{18}) [6,7]:

$$\log_{10}[1078 - T_b(\text{K})] = 3.03191 - 0.0499901\, N_C^{2/3} \qquad (9.2.1)$$

T_b values of perfluorinated *n*-alkanes have been fitted into the following model (C_1–C_{16}) [8]:

$$T_b(\text{K}) = 540.87 \log_{10}(N_C + 3) - 183.67 \qquad (9.2.2)$$

A compilation of structure–T_b relationships for homologous series has been given by Horvath [9]. This author also reviews various other structure–T_b relationships. Most of the available methods are restricted to classes of certain hydrocarbons or monofunctional derivatives thereof. In the following, models have been selected in which different molecular descriptors are employed to estimate T_b.

Correlation of Seybold   Seybold et al. [10] derived the following correlation for *n*-alkanes (C_2–C_8) and their branched isomers:

$$T_b(^\circ\text{C}) = -126.19 + 33.42\, N_C - 6.286\, N_{CH_3} \qquad n = 39, \quad s = 5.86, \quad r^2 = 0.987 \qquad (9.2.3)$$

For alkanols (C_1–C_{10}), expanding on model (8.4.1) by introducing N_{C_α}, the number of carbons bonded to the alpha carbon, yields [10]:

$$T_b(^\circ\text{C}) = 54.93 + 19.55 N_C - 7.52 N_{CH_3} + 8.99 N_{C_\alpha} \qquad n = 37, \quad s = 1.44,$$
$$r^2 = 0.993 \qquad (9.2.4)$$

TABLE 9.2.1   Parameters for Eq. 9.2.5 [11]

Compound Class	$a_0(°C)$	$a_1(°C)$	n	s	r	F
Alkyl halides	-108.431	226.874	24	16.35	0.896	$F(1,22) = 89.6$
Alkanols	5.019	127.969	48	8.25	0.964	$F(1,46) = 605.24$
Monoalkyl amines	-60.175	166.419	21	5.128	0.995	$F(1,19) = 2060.80$
Dialkyl amines	-71.007	157.702	13	4.100	0.997	$F(1,22) = 1954.29$

Van der Waals Volume–Boiling Point Relationships   Bhatnagar et al. [11] have found significant correlations between T_b and V_{vdW}:

$$T_b(°C) = a_0 - a_1 V_{vdW} \qquad (9.2.5)$$

where a_0 and a_1 are empirical, compound-class specific constants. They are listed in Table 9.2.1 for alkyl halides ($C_2–C_5$, F, Cl, Br, I), alkanols ($C_4–C_9$), monoalkyl amines ($C_3–C_{11}$) and dialkyl amines ($C_3–C_{10}$) along with the statistical parameters. The molecular descriptor, V_{vdW}, discriminates between isomers in certain cases, but not all. For example, V_{vdW} is 155.8 Å^3 for 1-nonanol and 150.8 Å^3 for 2-, 3-, 4-, and 5-nonanol.

Geometric Volume–Boiling Point Relationships   Bhattacharjee and Dasgupta [12,13] introduced the geometric volume, V_g, as molecular descriptor for alkanes, halomethanes, and haloethanes. A bilinear relationship has been reported for alkanes ($C_1–C_8$):

$$T_b(°C) = N_C(41.4974V_g - 80.877) + (-383.79V_g + 1242.733) \qquad (9.2.6)$$

where N_C is the number of carbon atoms per molecule and V_g is the geometric volume. This model accounts correctly for the increase of T_b with increasing V_g (parallel to the increase in molecular size) and the decrease of T_b with increasing V_g (parallel to the increase of branchedness) among isomers. For haloethanes, the following correlation has been derived:

$$
\begin{aligned}
T_b(°C) = {} & 1322.958 + [-287.168V_g + 88.37874V_g^2 - 8.41772V_g^3] \\
& + [-309.381V_{com} + 30.68011V_{com}^2 - 1.63856V_{com}^3] \\
& + [48.34256N_H + 11.42358N_H^2 - 1.76873N_H^3] \\
& n = 41, \quad s = 20.43, \quad r^2 = 0.9472
\end{aligned}
\qquad (9.2.7)
$$

where V_g is the geometric volume in Å^3, V_{com} is the "common" volume in Å^3, and N_H is the number of hydrogen atoms per molecule. Equation (9.2.5) has been applied to the computation of T_b for all 629 haloethanes that are theoretically possible by different combinations of F, Cl, Br, and I substituents [14].

MCI–Boiling Point Relationships   Kier and Hall [15], using connectivity indices, reported the following fit for alkanes ($C_5–C_9$):

$$T_b(°C) = -96.13 + 55.69\,^1\chi + 4.708\,^4\chi_{pc} \qquad n = 51, \quad s = 2.53, \quad r^2 = 0.994$$

$$(9.2.8)$$

Needham, et al. [16] derived the following model for alkanes (C_2–C_9):

$$T_b(°C) = -9.6(\pm 4.1) + 38.1(\pm 1.0)\,^1\chi - 49.0(\pm 19.3)(^0\chi)^{-1} + 5.7(\pm 0.3)^4\chi_{pc}$$
$$- 94.5(\pm 9.8)\chi_t + 8.4(\pm 2.5)^6\chi_p \qquad n = 74, \quad s = 1.86, \quad r^2 = 0.999,$$
$$F = 9030 \tag{9.2.9}$$

White [17] has derived an univariate relationship for PAHs:

$$T_b(K) = 225.71 + 76.21\,^1\chi_v \qquad n = 30, \quad s = 8.59, \quad r = 0.994 \tag{9.2.10}$$

Correlation of Grigoras Grigoras [18] derived a multilinear correlation to estimate $T_b(K)$ for liquid compounds, including saturated, unsaturated, and aromatic hydrocarbons, alcohols, acids, esters, amines, and nitriles:

$$T_b(K) = (127.7 \pm 6.1) + (0.718 \pm 0.038)A - (1.015 \pm 0.030)A_+$$
$$+ (0.230 \pm 0.024)A_- + (8.800 \pm 0.225)A_{HB} \qquad n = 137, \quad s = 14.1,$$
$$r = 0.979, \quad F = 745.1 \tag{9.2.11}$$

where A is the total molecular surface area, A_+ the sum of the surface areas of positively charged atoms multiplied by their corresponding scaled net atomic charge, A_- the sum of the surface areas of negatively charged atoms multiplied by their corresponding scaled net atomic charge, and A_{HB} the sum of the surface areas of hydrogen-bonding hydrogen atoms multiplied by their corresponding scaled net atomic charge. A, A_+, A_- and A_{HB} are based on contact atomic radii [18].

Correlation of Stanton, Jurs, and Hicks Stanton et al. [19] have developed a combined model and separate models for furanes, tetrahydrofuranes (THFs), and thiophenes. Model development has been based on descriptor analysis with 209 training set compounds. A variety of different structural descriptors has been employed. A fit error of 4.9% for the combined data set, of 5.8% for the furan–THF subset, and of 3.8% for the thiophen subset has been reported for T_b.

Correlation of Wessel and Jurs Wessel and Jurs [20] have developed a six-parameter model to estimate T_b of hydrocarbons (C_2–C_{24}) including alkanes, alkenes, alkynes, cycloalkanes, alkyl-substituted cycloalkanes, benzenes and PAHs, and terpenes. The model is based on a training set of 300 compounds with T_b values ranging from 169.4 to 770.1 K, having an average computed error of approximately 4.4 K. The prediction set constituted of 56 compounds. A startup set of 81 descriptors was employed. Model derivation involved (1) descriptor ranking with Gram–Schmidt orthogonalization, and (2) leaps-and-bounds regression analysis. The final model is

$$T_b(K) = (-30.16 \pm 5.528) + (237.4 \pm 24.86)QNEG + (-0.2480 \pm 0.01663)DPSA$$
$$+ (114.2 \pm 20.70)FNSA + (0.8120 \pm 0.06976)ALLP2$$
$$+ (-19.29 \pm 0.9542)MOLC7 + (50.35 \pm 0.9302)M^{1/2} \qquad n = 296,$$
$$rms = 6.3\,K, \quad r = 0.997, \quad F = 9215.1 \tag{9.2.12}$$

where T_b is the normal boiling point, QNEG the charge on the most negative atom [21], DPSA the partial positive minus partial negative surface area [22], FNSA the fractional negative surface [22], ALLP2 the total paths per total number of atoms [23], MOLC7 the path cluster 3 molecular connectivity [24], and $M^{1/2}$ the square root of the molar mass.

Graph-Theoretical Indices – Boiling Point Relationships   Randic et al. [25] have compared several graph-theoretical descriptors and their use in correlation with boiling points of alkanes [25]. Schultz and Schultz [26] reported the following correlation for alkanes (C_2–C_{15}):

$$T_b(°C) = -256.96 + 190.94 \left(\log_{10}|\mathbf{A} + \mathbf{D}|\right)^{1/2} \qquad n = 41, \quad r^2 = 0.994 \quad (9.2.13)$$

Yang et al. [27] introduced the descriptors EA_Σ and EA_{max}, derived from the extended adjacency (*EA*) matrix. They report the following correlation for alkanes and alkanols:

$$\text{Alkanes}(C_2\text{–}C_{10}): \quad T_b(°C) = 134.79 + 3.45\,EA_\Sigma - 10.43\,EA_{max} \qquad n = 149,$$
$$s = 9.46, \quad r = 0.9914, \quad F = 4178.34 \qquad (9.2.14)$$
$$\text{Alkanols}(C_1\text{–}C_{10}): \quad T_b(°C) = 132.14 + 7.70\,EA_\Sigma - 26.36\,EA_{max} \qquad n = 37,$$
$$s = 6.35, \quad r = 0.9838, \quad F = 511.90 \qquad (9.2.15)$$

Using the charge index, J_2, the following correlation has been reported for alkanols (C_4–C_7) [28]:

$$T_b(°C) = 55.58 + 15.46 N_C - 103.32\,J_2 \qquad n = 29, \quad s = 12.16, \quad r = 0.956,$$
$$F = 138.8,\, p < 0.001 \qquad (9.2.16)$$

The correlation coefficient r increases from 0.705 for the univariate correlation between J_2 and N_C to 0.956 for this bivariate correlation.

For dialkyl ethers (C_3–C_{10}), the following model has been derived by Balaban et al. [29]:

$$T_b(°C) = 46.45(\pm 16.02) - 25.86(\pm 3.81)S_0 - 12.50(\pm 0.66)N_{CH_3}$$
$$+ 37.89(\pm 0.70)\,{}^0\chi^v \qquad n = 72, \quad s = 4.89, \quad r^2 = 0.987, \quad F = 1749$$
$$(9.2.17)$$

where S_0 is the electrotopological state for oxygen. Horvath reviews similar correlations between T_b and molecular connectivity indices for some other classes [9] and correlations between T_b and molecular weight for polyhalogenated methanes and ethanes [30]. Models to estimate T_b for diverse derivatives of heterocyclic compounds such as furan, tetrahydrofuran, and thiophene require a more diverse set of molecular descriptors [19]. Gálvez et al. [31] have designed new topological descriptors, the *charge indexes*, and reported their correlation with T_b of alkanes and alcohols.

9.3 GROUP CONTRIBUTION APPROACHES FOR T_b

The group contribution approach has been employed in different ways to model the relation between T_b and molecular structure:

- Additivity in polyhaloalkanes
- Additivity in rigid aromatics
- Indirect via T_c or Θ as "additive" parameter and use of eq. 9.1.1
- Indirect via V_c and P_c as "additive" parameters (Miller's method)
- Direct by using nonlinear GCM

Additivity in Polyhaloalkanes   A simple atom contribution model for polyhalomethanes has been reported [32]:

$$T_b(°C) = 18.73(\pm 3.46) - 38.08(\pm 1.09)N_F + 14.60(\pm 1.12)N_{Cl}$$
$$+ 42.18(\pm 1.21)N_{Br} + 75.78(\pm 1.81)N_I \quad n = 44, \quad s = 5.26,$$
$$r^2 = 0.99, \quad F = 2204 \tag{9.3.1}$$

For polyhalogentated n-alkanes, the following rule regarding interchange of halogen atoms has been given:

T_b increases by 45°C on replacing geminally one F atom in a fluorocarbon with Cl, by 75°C with Br, and by 115°C with I [32].	(R-9.3.1)

Replacement of one or two methylene groups in n-alkanes by an oxygen atom does not "appreciably" change T_b [32]. Differences of less than 15°C are observed for n-hexane, n-heptane, and n-octane.	(R-9.3.2)

Balaban et al. [33] studied the use of neural networks to establish relationships between halomethanes and atom contributions and between chlorofluorocarbons (C_1–C_4) and atom contributions. In addition to atom contribution, their relationships include molecular descriptors (i.e., the Wiener and J indices).

Additivity in Rigid Aromatic Compounds   Simamora et al. [34] have developed a GCM that applies to mono- and polycyclic rigid aromatic ring systems containing as substituents a single hydrogen-bonding group, (i.e., hydroxy, aldehydo, primary amino, carboxylic, or amide) as well as non-hydrogen-bonding groups (i.e., halo, methyl, cyano, and nitro groups). The method applies to homoaromatic and nitrogen-containing aromatic rings. The following formulas have been employed:

$$T_b(°C) = \tfrac{1}{21}\sum n_i b_i \quad n = 241, \quad s = 13.66, \quad r^2 = 0.9994 \tag{9.3.2}$$

where n_i is the number of occurrences of group i in the molecule and b_i is the contribution of group i. The method further employs two types of correction factors, designed as (1) intramolecular hydrogen-bonding parameters, and (2) biphenyl parameters.

Method of Hishino, Zhu, Nagahama, and Hirata (HZNH)   The method of Hishino et al. [35] can be used to estimate T_b at 1 atm ($=101.32\,kPa$) for mono-, di-, tri-, and tetra-substituted alkylbenzenes. Since this method allows calculation of the Antoine coefficients A, B, and C (see Section 7.3), estimation of T_b using eq. 9.4.1 at pressures in the region 1.33 to 199.98 kPa is possible.

Method of Joback and Reid   This model [36] has been based on a database of 438 organic liquids and yielded an average absolute error of 12.9 K, corresponding to a 3.6 average percent error using the T_b values of the training set. The GCM equation is:

$$T_b = 198.2 + \sum n_i(\Delta T_b)_i \qquad n = 438, \quad s = 17.9\,K \qquad (9.3.3)$$

where the summation is over all group types i. $(\Delta T_b)_i$ is the contribution for the ith group type and n_i is the number of times the group occurs in the molecule. Application of this model to pentachlorobenzene is demonstrated in Figure 9.3.1. The estimated normal boiling point is 297.5°C, compared to 277°C found in the literature [37].

Modified Joback Method   Devotta and Pendyala [38] have reported the inadequate accuracy for estimated T_b of aliphatic halogenated compounds using the method of Joback and Reid. They modified this method by providing contributions for fluorocarbon groups ($-CF_3$, $>CF_2$, and $>CF-$) and by additionally introducing correction terms for perhalogenation and partial halogenation. Their

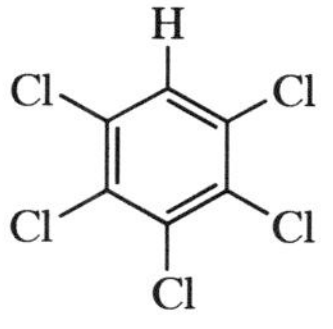

Pentachlorobenzene		
C_0 (eq. 9.3.3)	198.2	198.2
=CH–	1(26.73)	26.73
=C<	5(31.01)	155.05
–Cl	5(38.13)	190.65
	$T_b =$	570.63 K

$T_b = 297.5\,°C$ at 1 atm

Figure 9.3.1   Estimation of T_b (1 atm) for pentachlorobenzene using the method of Joback and Reid [36].

evaluation has been based on a set of 89 polyhalogenated alkanes and derivatives containing an ether, aldehyde, keto, or carboxylic acid, or amino group with T_b in the range 145 to 543 K. Application of their method has been demonstrated for tetrafluoromethane, 1,1,2,2-tetrafluoroethane, perfluorotrimethylamine, 1,1,1-trifluorochlorobromoethane, trifluorochloromethane, and 1,1,1-trichloroethane [38].

Method of Stein and Brown   The Stein and Brown model [39] is an extension of the method of Joback and Reid. By increasing the number of group types from 41 to 85, structurally broadened applicability and enhanced predictive accuracy has been gained. The model relies on a database of 4426 diverse organic liquids. It has been validated with 6584 other compounds, not used in the model derivation. Estimated T_b values had a average absolute error of 15.5 K, corresponding to a 3.2 average percent error for the training set, and an average absolute error of 20.4 K, corresponding to a 4.3 average percent error for the validation set. The additional groups in this model were derived by three different modifications:

1. Finer distinction with respect to structural environment
2. Combination of heteroatoms into larger functional units
3. Introduction of groups with B, Si, P, Se, and Sn atoms

Finer distinction, for example, has been derived for hydroxy groups, −OH. Joback and Reid distinguished only between aliphatic and phenolic −OH, whereas the new model distinguishes whether −OH is attached to a primary, secondary, tertiary, or aromatic C or non-C atom. The combination of heteroatoms into larger functional units refers to the definition of, for example, amido groups, −C(O)NH− and −C(O)N$\lessdot$, with individual contributions rather than adding up the contributions for the carbonyl and the amino group. New group contributions evaluated for groups such as $\gtrdot$Ph, $\gtrdot$SiH−, $\gtrdot$B−, −Se−, and $\gtrdot$Sn$\lessdot$ have been introduced. Application of this model to pentachlorobenzene is demonstrated in Figure 9.3.2. The estimated

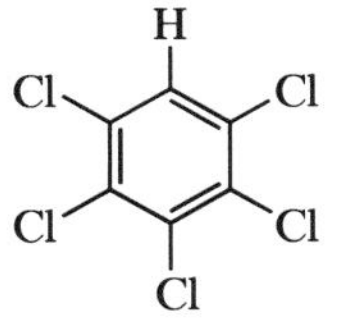

C_0 (eq. 9.3.3)	198.2	198.2
aaCH	1(28.53)	28.53
aaC−	5(30.76)	153.80
ϕ−Cl	5(36.79)	183.95
		———
	$T_b =$	564.48 K
	$T_b = 291.3\,°C$ at 1 atm	

Figure 9.3.2   Estimation of T_b (1 atm) for pentachlorobenzene using the method of Stein and Brown [39].

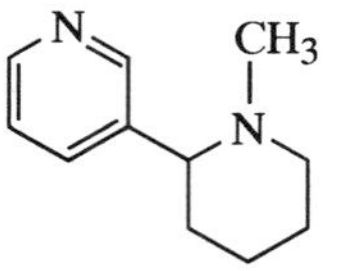

Nicotine

C_0 (eq. 9.3.3)	198.2	198.2
$-CH_3$	1(21.98)	21.98
$-C_rH_2-$	3(26.44)	79.32
$>C_rH-$	1(21.66)	21.66
aaCH	4(28.53)	114.12
aaC$-$	1(30.76)	30.76
$>N_r-$	1(32.77)	32.77
anN	1(39.88)	39.88
	$T_b =$	538.69 K

$$T_b = 265.5°C \text{ at } 1 \text{ atm}$$

Figure 9.3.3 Estimation of T_b (1 atm) for nicotine using the method of Stein and Brown [39].

normal boiling point is 291.3°C, an improvement over 297.5°C derived using the method of Joback and Reid (Figure 9.3.1), assuming that the experimental value is 277°C [37]. A second estimation example is shown in Figure 9.3.3 for nicotine. The method of Joback and Reid does not apply in this case because the contribution $>N-$ is available as nonring contribution only. Application of the method of Stein and Brown yields a value of 265.5°C for nicotine, which compares satisfactorily with the experimental value of 246.2°C [1].

Method of Wang, Milne, and Klopman The Wang et al. model [40] combines the approach of group contributions with local graph indices. A set of 49 contributions has been derived from a 541-compound database. The contributions are associated with either single- or multiatomic groups. For each group a molecule-specific group index, γ^G, is derived as the mean of the atomic γ values that apply to the atoms which are part of the particular group. The γ values are derived with the following equation:

$$\gamma_i = \log_{10} \frac{\sum(n_{ij})}{j+1} \tag{9.3.4}$$

where γ_i is the γ value of atom i in the molecule, n_{ij} the number of atoms at distance j from atom i, and the summation is carried over all distances j ranging from 1 to d_{max}. In this model, eq. 9.3.4 applies to the hydrogen-preserved molecular graph. The GCM equation is:

$$T_b(°K) = -33.02 + 37.47M^{0.5} + \sum[C_k P_k(1 - d_k \gamma_k^G)] \quad n = 541, \quad s = 7.57 \text{ K},$$
$$r^2 = 0.992, \quad F = 564.2 \tag{9.3.5}$$

where M is the molar mass, C_k the contribution of the kth group, P_k the number of occurrences of the kth group in the molecule, γ_k^G is γ^G for the kth group, d_k is the coefficient of γ_k^G, and the summation is carried over all groups in the molecule ($k = 1, 2, \ldots, 49$). The coefficients C_k and d_k are given in the source [40]. The prediction potential of this model has been examined by cross-validation tests.

Method of Lai, Chen, and Maddox The Lai et al. model [5] is a nonlinear GCM derived in a stepwise manner accounting for several functional groups in mono- and multifunctional compounds and for diverse factors such as branching, substitution and ring pattern, and hydrogen bonding. The approach is based on the following equation that applies for n-alkanes with a terminal function group:

$$T_b(\text{K}) = \left[\frac{a + b_C(1 - r_C)}{1 - r_C^{Nc}}\right] + \left[\frac{b_f + b_{fc}(1 - r_C)}{1 - r_C^{Nc}}\right] \qquad (9.3.6)$$

The left-hand term in eq. 9.3.6 corresponds to the n-alkyl contribution and the right-hand term to the functional group contribution. N_C is the number of carbon atoms in the molecule and r_C is a constant. The contribution parameters a and b_C refer to the alkyl group and the parameters b_f and b_{fc} to the functional group. For compounds with homogeneous multifunctional groups (e.g., alkanediols or polychlorinated alkanes), the model takes the following form:

$$T_b(\text{K}) = \left[\frac{a + b_C(1 - r_C)}{1 - r_C^{Nc}}\right] + \left[\frac{b_f + b_{fc}(1 - r_C)}{1 - r_C^{Nc}}\right][(1 - r_f)(1 - r_f^m)] \qquad (9.3.6a)$$

where m is the number of the particular function group in the molecule and r_f is a characteristic constant for the functional group. Modifying eq. 9.3.6a, the authors derived a general model for compounds with heterogeneous multifunctional groups (i.e., alkane molecules substituted by different groups). This model includes a term accounting for the interaction between different types of functional groups and has been further generalized by introducing structural corrections for the aforementioned factors. The authors employ 1169 compounds with known T_b to evaluate model accuracy and reliability. They demonstrate model application for 2'-methyl-1,1-diphenylethane and 4-chloro-2-methyl-2-butanol.

Method of Constantinou and Gani The Constantinou and Gani approach [41] is based on first- and second-order groups allowing a first-order approximation of T_b by solely using first-order groups and a more accurate estimations using groups of either order. The model is

$$\exp\left[T_b(K)/204.359\,\text{K}\right] = \sum n_i(T_{b1})_i + W\sum m_j(T_{b2})_j \qquad n = 392$$
$$W = 0: s = 10.48\,\text{K}, \quad \text{AAE} = 7.71\,\text{K}, \quad \text{AAPE} = 2.04\%$$
$$W = 1: s = 7.70\,\text{K}, \quad \text{AAE} = 5.35\,\text{K}, \quad \text{AAPE} = 1.42\%$$
$$(9.3.7)$$

where $(T_{b1})_i$ is the contribution of the first-order group type i, which occurs n_i times in the molecule and $(T_{b2})_j$ is the contribution of the second-order type j, with m_j occurrences in the molecule. W is zero or 1 for a first- and second-order approximation, respectively, and the statistical parameters are $s = [\sum (T_{b,\text{fit}} - T_{b,\text{obs}})^2 / n]^{1/2}$, $\text{AAE} = (1/n) \sum |T_{b,\text{fit}} - T_{b,\text{obs}}|$, and $\text{AAPE} = (1/n) \sum |T_{b,\text{fit}} - T_{b,\text{obs}}| / T_{b,\text{obs}} \times 100\%$.

Artificial Neural Network Model   Lee and Chen [42] have studied the ANN approach to design a GCM for the prediction of T_b, T_c, V_c, and the acentric factor of fluids. The network has a three-layer architecture. Input parameters are the numbers (per molecule) of 36 group types similar to those used in the method of Joback and Reid. The hidden layer contains three neurons, and the output layer four neurons, corresponding to the afore-listed properties. The sigmoid function has been selected as transfer function for each neuron. Weight adjustment has been derived by the back-propagation algorithm employing the generalized delta rule to minimize the mean-square error between desired and estimated property data. The average absolute deviations (AADs) of estimated from desired values for the ANN-based GCM has been compared with those for the conventional GCM of Joback and Reid. Significantly lower AADs have been found with the ANN model for all compound classes: namely, alkanes, alkenes, alkynes, alicyclics, aromatics, heterocycles, halocarbons, ethers/epoxides, esters, alcohols, aldehydes, acids, ketones, and amines/nitriles. The authors outline the superiority of the ANN model with built-in account for nonlinearity over the linear model according to eq. 1.6.3.

9.4   PRESSURE DEPENDENCE OF BOILING POINT

Rearrangement of the Antoine equation (7.4.1) leads to the following equation, which permits the estimation of boiling points from known Antoine constants within the applicable range for a given pressure:

$$T_b = \frac{B}{A - \log_{10} p_v} - C \qquad (9.4.1)$$

In Figure 9.4.1 we present the estimation of the normal boiling point for *n*-propylcyclopentane. The experimental reference is $T_b = 130.95°C$ [43]. In Figure 9.4.2 the estimation of the boiling point for 1-heptene at 737 mmHg is demonstrated. T_b (737 mmHg) $= 93.0°C$ is given in the literature [44].

Reduced-Pressure T_b–Structure Relationships   For certain compound classes, quantitative T_b–structure relationships are available to estimate T_b at reduced pressure. For example, the following equation has been reported by Kreglewski and Zwolinski for *n*-alkanes (C_6–C_{18}) [6], in analogy to eq. 9.2.1:

$$\log_{10}[989 - T_{b(50\,\text{mm})}(\text{K})] = 2.99615 - 0.0431882 N_C^{2/3} \qquad (9.4.2)$$

where $T_{b(50\,\text{mm})}$ is the boiling point at 50 mmHg.

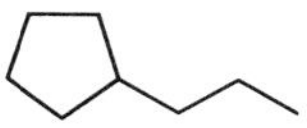

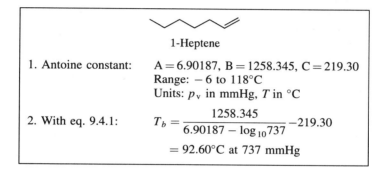

n-Propylcyclopentane

1. Antoine constant: A = 6.90392, B = 1384.386, C = 213.16
   Range: 21–158°C
   Units: p_v in mmHg, T in °C

2. With eq. 9.4.1: $T_b = \dfrac{1384.386}{6.90392 - \log_{10} 760} - 213.16$

   $= 131.0°C$

Figure 9.4.1 Estimation of T_b for *n*-propylcyclopentane using data from Dean [45].

1-Heptene

1. Antoine constant: A = 6.90187, B = 1258.345, C = 219.30
   Range: −6 to 118°C
   Units: p_v in mmHg, T in °C

2. With eq. 9.4.1: $T_b = \dfrac{1258.345}{6.90187 - \log_{10} 737} - 219.30$

   $= 92.60°C$ at 737 mmHg

Figure 9.4.2 Estimation of T_b (737 mmHg) for 1-heptene using data from Dean [45].

REFERENCES

1. Rechsteiner, C. E. , Boiling Point, in *Handbook of Chemical Property Estimation*, W. J. Lyman, W. F. Reehl, and D. H. Rosenblatt, Editors, 1990. Washington, DC: American Chemical Society.

2. Hagopian, J. H., Flash Points of Pure Substances, in *Handbook of Chemical Property Estimation*, W. J. Lyman, W. F. Reehl, and D. H. Rosenblatt, Editors, 1990, Washington, DC: American Chemical Society.

3. Satayanarayana, K., and P. G. Rao, Improved Equation to Estimate Flash Points of Organic Compounds. *J. Hazard. Mater.*, 1992: **32**, 81–85.

4. Vásquez, A., and J. G. Briano, Thermal Conductivity of Hydrocarbon Mixtures: A Perturbation Approach. *Ind. Eng. Chem. Res.*, 1993: **32**, 194–199.

5. Lai, W. Y., D. H. Chen, and R. N. Maddox, Application of a Nonlinear Group-Contribution Model to the Prediction of Physical Constants: 1. Predicting Normal Boiling Points with Molecular Structure. *Ind. Eng. Chem. Res.*, 1987: **26**, 1072–1079.

6. Kreglewski, A., and B. J. Zwolinski, A New Relation for Physical Properties of *n*-Alkanes and *n*-Alkyl Compounds. *J. Phys. Chem.*, 1961: **65**, 1050–1052.

7. Kudchadker, A. P., and B. J. Zwolinski, Vapor Pressures and Boiling Points of Normal Alkanes, C21 to C100. *J. Chem. Eng. Data*, 1966: **11**, 253–255.

8. Postelnek, W., Boiling Points of Normal Perfluoroalkanes. *J. Phys. Chem.*, 1959: **63**, 746–747.

9. Horvath, A. L., *Molecular Design: Chemical Structure Generation from the Properties of Pure Organic Compounds*, 1992. Amsterdam: Elsevier.

10. Seybold, P. G., M. May, and U. A. Bagal, Molecular Structure–Property Relationships. *J. Chem. Educ.*, 1987: **64**(7), 575–581.

11. Bhatnagar, R. P., P. Singh, and S. P. Gupta, Correlation of van der Waals Volume with Boiling Point, Solubility and Molar Refraction. *Indian J. Chem.*, 1980: **19B**, 780–783.

12. Bhattacharjee, S., and P. Dasgupta, Molecular Property Correlation in Haloethanes with Geometric Volume. *Comput. Chem.*, 1992: **16**, 223–228.

13. Bhattacharjee, S., and P. Dasgupta, Molecular Property Correlation in Alkanes with Geometric Volume. *Comput. Chem.*, 1994: **18**, 61–71.

14. Bhattacharjee, S., Haloethanes, Geometric Volume and Atomic Contribution Method. *Comput. Chem.*, 1994: **18**, 419–429.

15. Kier, L. B., and L. H. Hall, *Molecular Connectivity in Chemistry and Drug Research*, 1976. San Diego, CA: Academic Press.

16. Needham, D. E., I.-C. Wei, and P. G. Seybold, Molecular Modeling of the Physical Properties of the Alkanes. *J. Am. Chem. Soc.*, 1988: **110**, 4186–4194.

17. White, C. M., Prediction of the Boiling Point, Heat of Vaporization, and Vapor Pressure at Various Temperatures for Polycyclic Aromatic Hydrocarbons. *J. Chem. Eng. Data*, 1986: **31**, 198–203.

18. Grigoras, S., A Structural Approach to Calculate Physical Properties of Pure Organic Substances: The Critical Temperature, Critical Volume and Related Properties. *J. Comput. Chem.*, 1990: **11**, 493–510.

19. Stanton, D. T., P. C. Jurs, and M. G. Hicks, Computer-Assisted Prediction of Normal Boiling Points of Furans, Tetrahydrofurans, and Thiophenes. *J. Chem. Inf. Comput. Sci.*, 1991: **31**, 301–310.

20. Wessel, M. D., and P. C. Jurs, Prediction of Normal Boiling Points of Hydrocarbons from Molecular Structure. *J. Chem. Inf. Comput. Sci.*, 1995: **35**, 68–76.

21. Dixon, S. L., and P. C. Jurs, Atomic Charge Calculations for Quantitative Structure–Property Relationships. *J. Comput. Chem.*, 1992: **3**, 492.

22. Stanton, D. T., and P. C. Jurs, Development and Use of Charged Partial Surface Area Structural Descriptors for Quantitative Structure–Property Relationship Studies. *Anal. Chem.*, 1990: **62**, 2323–2329.

23. Wiener, H., Structural Determination of Paraffin Boiling Points. *J. Am. Chem. Soc.*, 1947: **69**, 17–20.

24. Randić, M., On Molecular Identification Numbers. *J. Chem. Inf. Comput. Sci.*, 1984: **24**, 164.

25. Randić, M., P. J. Hansen, and P. C. Jurs, Search for Useful Graph Theoretical Invariants of Molecular Structure. *J. Chem. Inf. Comput. Sci.*, 1988: **28**, 60–68.

26. Schultz, H. P., and E. B. Schultz, Topological Organic Chemistry: 2. Graph Theory, Matrix Determinants and Eigenvalues, and Topological Indices of Alkanes. *J. Chem. Inf. Comput. Sci.*, 1990: **30**, 27–29.

27. Yang, Y.-Q., L. Xu, and C.-Y. Hu, Extended Adjacency Matrix Indices and Their Applications. *J. Chem. Inf. Comput. Sci.*, 1994: **34**, 1140–1145.

28. Gálvez, J., et al., Charge Indexes: New Topological Descriptors. *J. Chem. Inf. Comput. Sci.*, 1994: **34**, 520–525.

29. Balaban, A. T., L. B. Kier, and L. H. Hall, Correlations Between Chemical Structure and Normal Boiling Points of Acyclic Ethers, Peroxides, Acetals, and Their Sulfur Analogues. *J. Chem. Inf. Comput. Sci.*, 1992: **32**, 237–244.

30. Horvath, A. L., Estimate Properties of Organic Compounds: Simple Polynomial Equations Relate the Properties of Organic Compounds to Their Chemical Structure. *Chem. Eng.*, 1988: Aug. 15, **95**(11), 155–158.

31. Gálvez, J., et al., Charge Indexes: New Topological Descriptors. *J. Chem. Inf. Comput. Sci.*, 1994: **34**, 520–525.

32. Balaban, A. T., et al., Correlations Between Chemical Structure and Normal Boiling Points of Halogenated Alkanes C1–C4. *J. Chem. Inf. Comput. Sci.*, 1992: **32**, 233–237.

33. Balaban, A. T., et al., Correlation Between Structure and Normal Boiling Points of Haloalkanes C1–C4 Using Neural Networks. *J. Chem. Inf. Comput. Sci.*, 1994: **34**, 1118–1121.

34. Simamora, P., A. H. Miller, and S. H. Yalkowsky, Melting Point and Normal Boiling Point Correlations: Applications to Rigid Aromatic Compounds. *J. Chem. Inf. Comput. Sci.*, 1993: **33**, 437–440.

35. Hishino, D., et al., Prediction of Vapor Pressures for Substituted Benzenes by a Group-Contribution Method. *Ind. Eng. Chem. Fundam.*, 1985: **24**, 112–114.

36. Joback, K. G., and R. C. Reid, Estimation of Pure-Component Properties from Group-Contribution. *Chem. Eng. Commun.*, 1987: **57**, 233–243.

37. Miller, M. M., et al., Aqueous Solubilities, Octanol/Water Partition Coefficients, and Entropies of Melting of Chlorinated Benzenes and Biphenyls. *J. Chem. Eng. Data.*, 1984: **29**, 184–190.

38. Devotta, S., and V. R. Pendyala, Modified Joback Group Contribution Method for Normal Boiling Point of Aliphatic Halogenated Compounds. *Ind. Eng. Chem. Res.*, 1992: **31**, 2042–2046.

39. Stein, S. E., and R. L. Brown, Estimation of Normal Boiling Points from Group Contributions. *J. Chem. Inf. Comput. Sci.*, 1994: **34**, 581–587.

40. Wang, S., G. W. A. Milne, and G. Klopman, Graph Theory and Group Contributions in the Estimation of Boiling Points. *J. Chem. Inf. Comput. Sci.*, 1994: **34**, 1242–1250.

41. Constantinou, L., and R. Gani, New Group Contribution Method for Estimating Properties of Pure Compounds. *AIChE J.*, 1994. **40**, 1697–1710.

42. Lee, M. J., and J.-T. Chen, Fluid Property Predictions with the Aid of Neural Networks. *Ind. Eng. Chem. Res.*, 1993: **32**, 995–997.

43. Forziati, A. F., and F. D. Rossini, Physical Properties of Sixty API–NBS Hydrocarbons. *J. Res. Nat. Bur. Stand.*, 1949: **43**, 473–476.

44. Campbell, K. N., and L. T. Eby, The Reduction of Multiple Carbon–Carbon Bonds: III. Further Studies on the Preparation of Olefins from Acetylenes. *J. Am. Chem. Soc.*, 1941: **63**, 2683–2685.

45. Dean, J. A., *Lange's Handbook of Chemistry*, 14th ed., 1992. New York: McGraw-Hill.

CHAPTER 10

MELTING POINT

10.1 DEFINITIONS AND APPLICATIONS

The melting point of a compound, T_m, is the temperature at which the transition from the solid phase into the liquid phases takes place for a given pressure. At the melting point, the solid phase coexists in equilibrium with the liquid phase. The melting point at 1 atm is occasionally referred to as the *normal melting point* (compare with *normal boiling point*). However, literature references to the melting point in most cases mean, by default, the normal melting point.

The term *melting point* is frequently used interchangeably with the term *freezing point*. The difference between the two is the direction of approach to equilibrium. For a one-component system, these two points coincide; for complex systems they generally differ [1]. For certain compounds a melting point might not be measurable because the compound, exposed to temperature increase, undergoes a chemical reaction before the melting process can occur.

USES FOR MELTING POINT DATA

- To indicate (together with the boiling point) the physical state of a compound
- To assess the purity of a compound
- To estimate the surface tension (eq. 5.2.1)
- To estimate the vapor pressure with QPPRs (eqs. 7.2.1, 7.2.2, and 7.2.3)
- To estimate aqueous solubility of solids using QPPRs (Section 11.4; eq. 11.7.7)
- To estimate the *n*-octanol/water partition coefficient using QPPRs (Section 13.2)

It is justified to say that there are many more compounds with data known for the melting point than probably for any other measurable compound property. Despite this magnificent pool of data as a potential evaluation set to design structure–T_m relationships, the number of such relationships that are applicable to the accurate estimation of T_m, is extremely low. For example, Needham et al. [2] found that correlations between T_m of alkanes and molecular descriptors showed unsatisfactory statistics, whereas analogous correlations for T_b, T_c, P_c, V_M, R_D, ΔH_v, and σ gave excellent statistical results. An explanation for the lack of applicable structure–T_m correlations is the strong dependence of T_m on the three-dimensional structure of the solid state (i.e., the molecular arrangement in crystal states and the significance of intermolecular bonding). The following facts make systematic study of structure–T_m correlations difficult:

- Multiple melting points due to different solid-phase modifications
- Existence of one or more liquid crystal phases
- Occurrence of chemical transformations (rearrangement, decomposition, poly-merization)

Multiple Melting Points A compound may have different crystal structures (i.e., solid phases). For example, carbon tetrachloride has three known solid phases at atmospheric pressure: Ia (face-centered cubic), Ib (rhombohedral), and II (monoclinic). Ia and Ib melt at temperatures some 5 K apart [3]. Multiple melting points have been reported for a large set of compounds, such as many of those listed in the *Merck Index* [4]. Dearden and Rahman "improved" a structure–melting point correlation for substituted anilines by excluding two outliers on the ground that their T_m values were inadequate, due to different crystalline forms [5].

Liquid Crystals Liquid-crystal phases may occur between the solid and the liquid phase. Cholesteryl myristate, for example, exists in a liquid-crystal phase between 71 and 85°C [6]. The appearance of liquid-crystal phases depends on the molecular structure. Compounds with elongated structures that are fairly rigid in the central part of the molecule are likely candidates for liquid crystals. The homologous series of *p*-alkoxybenzylidene-*p*-*n*-butylanilines is just one example for compounds with liquid-crystal phases. An excellent introduction to liquid crystals and their properties has been written by Collings [6].

Estimation of Melting Points As indicated above, the development of structure–T_m relationships is not as straightforward as it is for other properties. In the following sections we discuss briefly the estimation of T_m for homologous series and for other sets of structurally related compounds. A GCM designed to estimate T_m for more diverse sets of compounds is introduced. Although not very accurate, the GCM approach may be applicable for the following tasks:

- To decide if a compound is in the solid or fluid phase at a given temperature
- To estimate T_m for a compound if T_m is known for structurally related compounds

Both cases are illustrated in Section 10.4 with a variety of examples.

10.2 HOMOLOGOUS SERIES AND T_m

For homologous series, correlations between T_m and N_{CH_2} depend on whether N_C is odd or even. The odd–even effect has been discussed in Section 1.3 and elsewhere [7,8]. For alkanes it vanishes above $N_{CH_2} = 30$. Then the melting points fall on a smooth curve where T_m increases with increasing N_{CH_2} toward an upper limit given by the melting point of polyethylene: $T_m^\infty = 414.6\,K$ [9]. Relationships between T_m and N_C have been studied for various homologous series (see odd–even effect in Section 1.3). Somayajulu [9] has reported the following relationship for homologous series of the general formula $Y-(CH_2)_k-H$:

$$\ln[(T_m^\infty - T_m)(K)] = a - b(k)^{1/25} \qquad (10.2.1)$$

TABLE 10.2.1 Values of the Parameters in Eq. 10.2.1 for Selected Homologous Series

Homologous Series	a	b	k^*	s^a
n-Alkanes	24.71207	17.79905	31	0.947
Cycloalkanes	30.35974	22.57216	31	1.92
1-Alkylcyclopentanes	27.16582	19.80791	22	0.341
1-Alkylcyclohexanes	28.58733	21.11261	25	0.438
1-Alkenes	29.29506	19.13557	21	0.009
1-Alkynes	26.42416	19.32058	15	0.637
1-Alkylbenzenes	28.71740	21.18813	16	0.148
1-Alkylnaphthalenes	25.15359	18.06739	25	—
2-Alkylnaphthalenes	26.00394	18.80971	25	—
1-Fluoroalkanes	26.55369	19.44985	30	—
1-Chloroalkanes	25.67164	18.64411	30	—
1-Bromoalkanes	24.48168	17.59152	22	0.168
1-Iodoalkanes	22.55096	15.95350	30	1.29
1-Alkanols	24.11107	17.55276	21	0.736
2-Alkanols	24.26195	17.64788	30	—
n-Alkanoic acids	20.89539	14.85653	22	1.10
1-Alkanals	26.25112	19.17364	30	—
2-Alkanones	23.80299	17.20223	15	0.278
Methyl alkanoates	26.62865	19.52636	22	1.85
Ethyl alkanoates	30.02291	22.44980	22	1.17
n-Alkyl methanoates	25.67164	18.64411	28	—
n-Alkyl ethanoates	27.71664	20.42408	20	1.89
Dialkyl ethers	24.56745	17.18798	28	—
1-Alkanethiols	25.39403	18.39017	30	—
2-Alkanethiols	24.86143	17.90249	30	—
2-Thioalkanes	24.60585	17.66823	30	—
1-Alkanamines	22.85642	16.32426	22	—
Dialkyl amines	24.67382	17.72836	28	—
Trialkyl amines	26.84949	19.07693	36	0.94
1-Alkanenitriles	26.55369	19.44985	30	—

[a] Standard deviation (not shown when graphically extrapolated T_m values have been used).
Source: Ref. 9. Reprinted with permission. Copyright © 1990 Plenum Publishing Corp.

where T_m^∞ is 414.6 K and a and b are compound class specific parameters and k is the chain length . Equation 10.2.1 is applicable above a given k^*, depending on the functional group Y. Below k^* the odd–even effect has to be considered. Note that k differs from N_{CH_2} in all cases where Y also contains CH_2 groups (e.g., in the series of 1-alkylcyclopentanes). The coefficients a and b for various homologous series along with their k^* values are listed in Table 10.2.1.

10.3   GROUP CONTRIBUTION APPROACH FOR T_m

The GCM approach has been applied to the estimation of T_m for organic compounds containing functional groups with O, S, N, and halogen atoms [10], for rigid aromatic compounds [11], and for organic polymers with various possible substituents [12]. The latter method employs various corrections that account for special structural features in the polymer molecule. The first two methods are described below.

Method of Simamora, Miller, and Yalkowsky   The Simamora et al. method [11] has been developed for mono- and polycyclic rigid aromatic ring systems containing as substituents a single hydrogen-bonding group (i.e., hydroxyl, aldehydo, primary amino, carboxylic, or amide as well as non-hydrogen-bonding groups) (i.e., halo, methyl, cyano, and nitro groups). The method applies to homoaromatic and nitrogen-containing aromatic rings. The model equations is as follows::

$$T_m(°C) = \frac{1}{13.5 - 4.6 \log \sigma} \sum n_i m_i \qquad n = 1181, \quad s = 36.63, \quad r^2 = 0.9910$$

$$(10.3.1)$$

where n_i is the number of occurrences of group i in the molecule, m_i is the contribution of group i, and σ is the rotational symmetry, defined as the number of ways that a molecule can rotate to give indistinguishable images. The method further employs two types of correction factors, designed as (1) intramolecular hydrogen bonding parameters, and (2) biphenyl parameters.

Method of Constantinou and Gani   The Constantinou and Gani approach [13] has been described for T_b in Section 9.3. The analog model for T_m is

$$\exp[T_m(K)/102.425 \text{ K}] = \sum n_i(T_{m1})_i + W \sum m_j(T_{m2})_j \qquad n = 312$$

$$W = 0 : s = 22.51 \text{ K}, \quad \text{AAE} = 17.39 \text{ K}, \quad \text{AAPE} = 8.90\%$$

$$W = 1 : s = 18.28 \text{ K}, \quad \text{AAE} = 14.03 \text{ K}, \quad \text{AAPE} = 7.23\%$$

$$(10.3.2)$$

where $(T_{m1})_i$ is the contribution of the first-order group type i which occurs n_i times in the molecule, and $(T_{m2})_j$ is the contribution of the second-order type j, with m_j occurrences in the molecule. W is zero or 1 for a first- or second-order approximation, respectively and the statistical parameters are $s = [\sum(T_{b,\text{fit}} - T_{b,\text{obs}})^2/n]^{1/2}$, $\text{AAE} = (1/n) \sum |T_{b,\text{fit}} - T_{b,\text{obs}}|$, and $\text{AAPE} = (1/n) \sum |T_{b,\text{fit}} - T_{b,\text{obs}}|/T_{b,\text{obs}} \times 100\%$.

Method of Joback and Reid   The Joback and Reid model [10] has been based on a database of 388 organic compounds and yielded an average absolute error of 22.6 K, corresponding to a 11.2 average percent error for the retro-estimated T_m values of the training set. The GCM equation is

$$T_m(K) = 122.5 + \sum n_i(\Delta T_m)_i \qquad n = 388, \quad s = 24.7\,K \qquad (10.3.3)$$

where the summation is over all group types i. $(\Delta T_m)_i$ is the contribution for the ith group type and n_i is the number of times the group occurs in the molecule. Application of this model to cyclopropyl methyl ether, 1,2-cyclopentenophenanthrene, and anethol is demonstrated in Figures 10.3.1 to 10.3.3. The corresponding estimated melting points are $-110, 160$, and $-14.6°C$. Experimental data are -119, $135-136$, and $21.4°C$ [4], respectively.

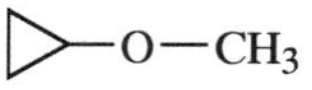

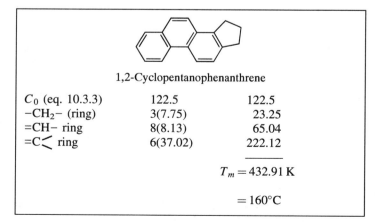

Cyclopropyl methyl ether		
C_0 (eq. 10.3.3)	122.5	122.5
$-CH_3$	1(-5.10)	-5.10
$-CH_2-$ (ring)	2(7.75)	15.50
$\gtrdot CH-$ (ring)	1(8.13)	8.13
$-O-$	1(22.23)	22.23
		$T_m = 163.26\,K$
		$= -110°C$

Figure 10.3.1   Estimation of T_m for cyclopropyl methyl ether using the method of Joback and Reid [10].

1,2-Cyclopentanophenanthrene		
C_0 (eq. 10.3.3)	122.5	122.5
$-CH_2-$ (ring)	3(7.75)	23.25
$=CH-$ ring	8(8.13)	65.04
$=C\lneqq$ ring	6(37.02)	222.12
		$T_m = 432.91\,K$
		$= 160°C$

Figure 10.3.2   Estimation of T_m for 1,2-cyclopentanophenanthrene using the method of Joback and Reid [10].

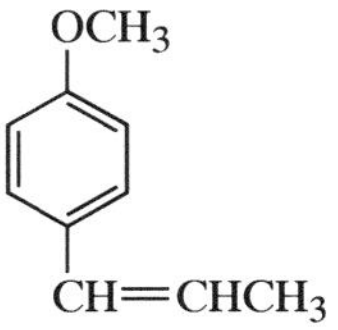

OCH$_3$

CH=CHCH$_3$

Anethol

C_0 (eq. 10.3.3)	122.5	122.5
–CH$_3$	2(– 5.10)	– 10.20
=CH–	2(8.73)	17.46
=CH– ring	4(8.13)	32.52
=C< ring	2(37.02)	74.04
–O–	1(22.23)	22.23

$T_m = 258.55$ K

$= – 14.6°C$

Figure 10.3.3   Estimation of T_m for anethol using the method of Joback and Reid [10].

Suppose that the compound's phase at 25°C was of interest. This question would have been answered correctly for all three compounds, although the quantitative estimation of T_m is not very precise. Suppose that the compound's phase at 20°C was of interest. This question would have been answered correctly for cyclopropyl methyl ether and 1,2-cyclopentenophenanthrene, but not for anethol. The magnitude of the interval $|(T_m)_{estimated} - T_{interest}|$ can serve as a confidence measure for binary decision of the foregoing type. If $|(T_m)_{estimated} - T_{interest}|$ is lower than 50°C, a decision as to whether a compound is fluid or solid at $T_{interest}$ should not be made based on T_m estimated using the method of Joback and Reid.

10.4.   ESTIMATION OF T_m BASED ON MOLECULAR SIMILARITY

Structurally similar compounds often exhibit large differences in their melting points. This is illustrated in Figure 10.4.1 by comparing T_m of aromatic aldehydes and analogous carboxylic acid compounds. Structurally, the compounds differs by merely

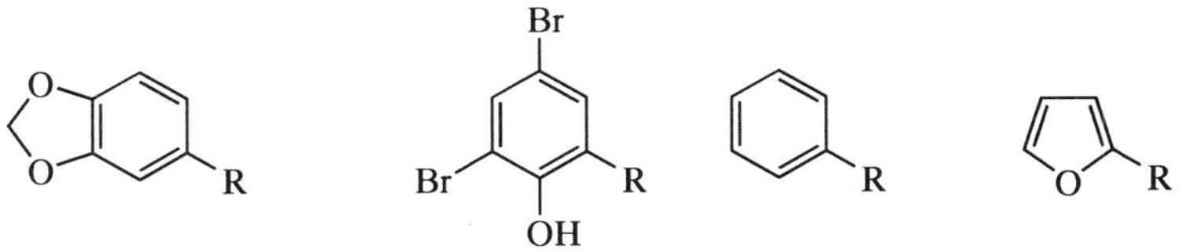

R = –C(=O)H $T_m = 37°C$	$T_m = 86°C$	$T_m = -56.5°C$	$T_m = -36.5°C$
R = –C(=O)OH $T_m = 229°C$	$T_m = 223°C$	$T_m = 122.4°C$	$T_m = 133-134°C$

Figure 10.4.1 T_m for aromatic aldehydes and their analogous carboxylic acid compounds [4].

3-Amino-2-naphthoic acid

T_m = 214°C [4]

3-Amino-2-naphthoic acid ethyl ester

T_m = ?

Deletion		
–COOH (acid)	1(155.50)	155.50
	Δ_{DE} =	155.50
Insertion		
–COO– (ester)	1(53.60)	53.60
$-CH_2-$	1(11.27)	11.27
$-CH_3$	1(–5.10)	–5.10
	Δ_{IN} =	59.77

$$T_m = 214 - 155.50 + 59.77 = 118.27°C$$
$$= 118°C$$

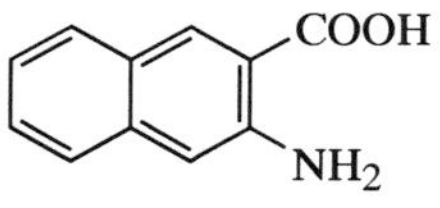

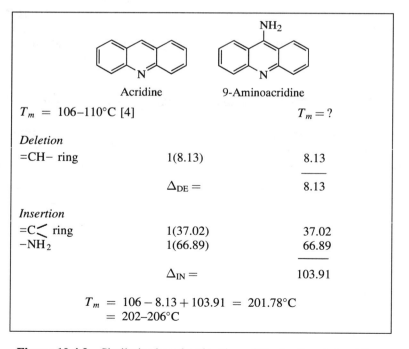

Figure 10.4.2 Similarity-based estimation of T_m for 3-amino-2-naphthoic acid ethyl ester.

Acridine

9-Aminoacridine

T_m = 106–110°C [4] T_m = ?

Deletion		
=CH– ring	1(8.13)	8.13
	Δ_{DE} =	8.13
Insertion		
=C< ring	1(37.02)	37.02
$-NH_2$	1(66.89)	66.89
	Δ_{IN} =	103.91

$$T_m = 106 - 8.13 + 103.91 = 201.78°C$$
$$= 202\text{–}206°C$$

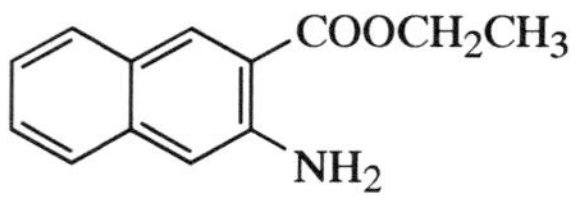

Figure 10.4.3 Similarity-based estimation of T_m for 9-aminoacridine.

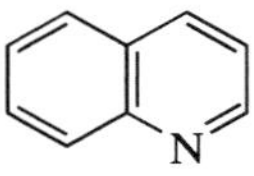

Quinoline		Quininic acid
$T_m = -15°C$ [4]		$T_m = ?$

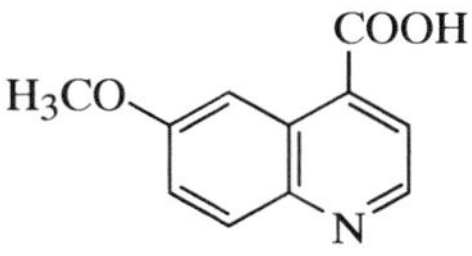

Deletion

=CH– ring	2(8.13)	16.26
		————
	$\Delta_{DE} =$	16.26

Insertion

=C< ring	2(37.02)	74.04
–CH$_3$	1(– 5.10)	– 5.10
–O– (nonring)	1(22.23)	22.23
–COOH	1(155.50)	155.50
		————
	$\Delta_{IN} =$	246.67

$$T_m = -15 - 16.26 + 246.67 = 215.41°C$$
$$= 215°C$$

Figure 10.4.4 Similarity-based estimation of T_m for quininic acid.

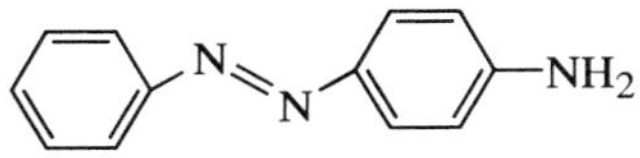

p-Aminoazobenzene		*o*-Aminoazotoluene
$T_m = 128°C$ [4]		$T_m = ?$

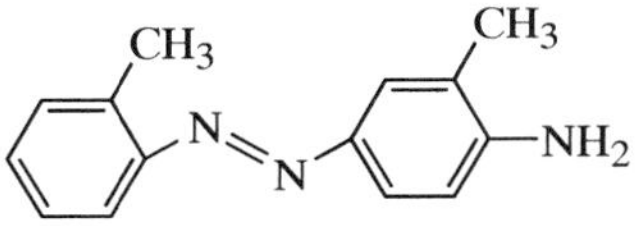

Deletion

=CH– ring	2(8.13)	16.26
		————
	$\Delta_{DE} =$	16.26

Insertion

=C< ring	2(37.02)	74.04
–CH$_3$	2(– 5.10)	– 10.20
		————
	$\Delta_{IN} =$	63.84

$$T_m = 128 - 16.26 + 63.84$$
$$= 175.58°C$$

Figure 10.4.5 Similarity-based estimation of T_m for *o*-aminoazotoluene.

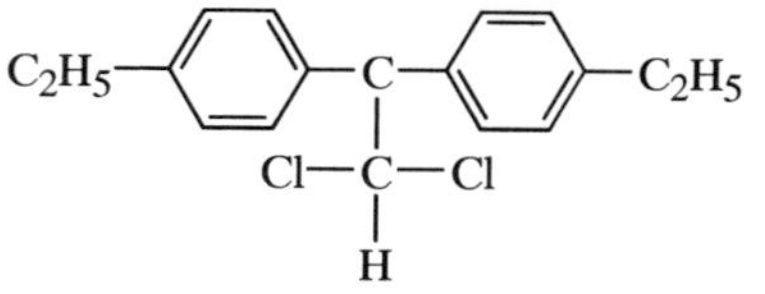

1,1-Dichloro-2,2-bis(*p*-ethylphenyl)ethane

T_m = 56–57°C [4]

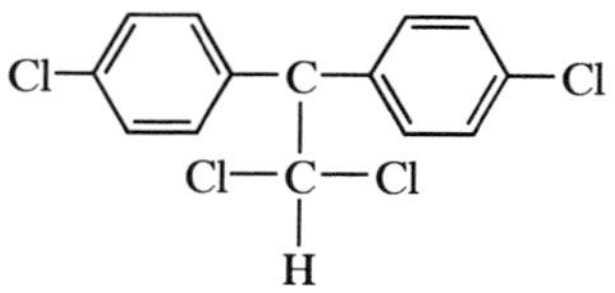

1,1-Dichloro-2,2-bis(*p*-chlorophenyl)ethane

T_m = ?

Deletion		
$-CH_3$	2(−5.10)	−10.20
$-CH_2-$	2(11.27)	22.54
	$\Delta_{DE} =$	12.34
Insertion		
−Cl	2(13.55)	27.10
	$\Delta_{IN} =$	27.10

$$T_m = 56 - 12.34 + 27.10 = 70.76°C$$
$$= 71°C$$

Figure 10.4.6 Similarity-based estimation of T_m for 1,1-dichloro-2,2-bis(*p*-chlorophenyl)-ethane.

one O atom inserted between an aldehyde H atom and a C atom. The mean T_m difference for the four compound pairs is 169.5°C. For the same structural difference, a T_m difference of 96.85°C is derived using the GCM of Joback and Reid. Note that this GCM does not distinguish between aliphatic and aromatic aldehyde and carboxylic groups.

Clearly, this example demonstrates how important it is to recognize the structural difference between similar compounds and base property estimation on ΔStructure–ΔT_m relationships instead of simply setting their T_m values equal to each other. Figures 10.4.2 to 10.4.6 illustrate similarity-based estimation of T_m using the method of Joback and Reid (Section 9.3). For comparison, the observed T_m values [4] for the query compounds are given below:

3-Amino-2-naphthoic acid ethyl ester:	$T_m = 115-115.5°C$
9-Aminoacridine:	$T_m = 241°C$
Quininic acid:	$T_m \approx 280°C$(decomposition)
o-Aminoazotoluene:	$T_m = 101-102°C$
1, 1-Dichloro-2, 2-bis(*p*-chlorophenyl)ethane:	$T_m = 109-110°C$

REFERENCES

1. Horvath, A. L., *Molecular Design: Chemical Structure Generation from the Properties of Pure Organic Compounds*, 1992. Amsterdam: Elsevier.
2. Needham, D. E., I.-C. Wei, and P. G. Seybold. Molecular Modeling of the Physical Properties of the Alkanes. *J. Am. Chem. Soc.*, 1998: **110**, 4186–4194.

3. Bean, V. E., and S. D. Wood, The Dual Melting Curves and Metastability of Carbon Tetrachloride. *J. Chem. Phys.*, 1980: **72**, 5838–5841.

4. Merck, *The Merck Index: An Encyclopedia of Chemicals, Drugs, and Biologicals*, 11th ed., 1989. Rahway, NJ: Merck and Co., Inc.

5. Dearden, J. C., and M. H. Rahman, QSAR Approach to the Prediction of Melting Points of Substituted Anilines. *Math. Comput. Model.*, 1988: **11**, 843–846.

6. Collings, P. J., *Liquid Crystals*, 1990. Princeton, NJ: Princeton University Press.

7. Burrows, H. D., Studying Odd–Even Effects and Solubility Behavior Using α,ω-Dicarboxylic Acids. *J. Chem. Educ.*, 1992: **69**, 69–73.

8. Francis, F., and S. H. Piper, The Higher *n*-Aliphatic Acids and Their Methyl and Ethyl Esters. *J. Am. Chem. Soc.*, 1939: **61**, 577–581.

9. Somayajulu, G. R., The Melting Point of Ultralong Paraffins and Their Homologues. *Int. J. Thermophys.*, 1990: **11**, 555–572.

10. Joback, K. G., and R. C. Reid, Estimation of Pure-Component Properties from Group-Contribution. *Chem. Eng. Commum.*, 1987: **57**, 233–243.

11. Simamora, P., A. H. Miller, and S. H. Yalkowsky, Melting Point and Normal Boiling Point Correlations: Applications to Rigid Aromatic Compounds. *J. Chem. Inf. Comput. Sci.*, 1993: **33**, 437–440.

12. van Krevelen, D. W., Properties of Polymers. 3rd ed., 1990. Amsterdam: Elsevier.

13. Constantinou, L., and R. Gani, New Group Contribution Method for Estimating Properties of Pure Compounds. *AIChE J.*, 1994: **40**, 1697–1710.

CHAPTER 11

AQUEOUS SOLUBILITY

11.1 DEFINITION

Water solubility is defined as the saturation concentration of a compound in water, that is the maximum amount of the compound dissolved in water under equilibrium conditions. The most common units used to express water solubility are

- *Mass-per-volume water solubilities*, C_w, are given in units of $mol\,L^{-1}$ or $g\,L^{-1}$, stating the amount of solute per liter of solution.
- *Mass-per-mass water solubilities*, S_w, have been reported in units of g/g% (i.e., grams of compound per hundred grams of water). The units ppmw (parts per million on weight basis) or ppbw (parts per billion on weight basis) are also commonly used.
- *Mole fraction water solubilities*, X_k, are conveniently used in solubility–temperature and in multicomponent representations of solubility information. The mole fraction, X_k, of a component k in a system of m components is defined as

$$X_k = \frac{n_k}{\sum n_i} \tag{11.1.1}$$

where n_i is the number of moles of component i and the summation is from $i = 1$ to m. For example, X_{HC} and X_w denote the mole fraction for a hydrocarbon solved in water and for water solved in the hydrocarbon, respectively. The mole fraction solubility at saturation is usually represented by the superscript s (in our example, X_{HC}^s and X_w^s). Note that the subscript in X_k indicates the solutes, whereas the subscripts in C_w and S_w state the type of solvent. In some cases the notation $X_{w,s}$ is used for the mole fraction of binary water–organic compound systems, where the subscripts w and s refer to water and the organic substance, respectively.

The units for the solubilities defined above are not interconvertible, unless further property data, such as the solution density, are known. Only for low concentrations can it be assumed that these solubilities are approximately proportional to each other [1].

Unit Conversion for Low Concentration Solubilities Mass-per-volume and mass-per-mass solubilities are related by

$$\text{For ``poorly'' soluble compounds:} \quad 1\,\text{g L}^{-1} = 1\,\text{ppmw} = 0.1\,\text{g/g\%}$$

if one assumes that the solvent and solution densities are equal. The relation between molar and mole fraction solubility is [2]

$$C_w(\text{mol L}^{-1}) = \frac{1000\rho_{\text{soln}}X_s^s}{18 + (M_s - 18)X_s^s} \tag{11.1.2}$$

where ρ_{soln} is the density of the saturated solution in g cm^{-1}, X_s^s is the mole fraction for the solute in its saturated solution with water, and M_s is the molecular mass of the solute. As X_s^s approaches zero and ρ_{soln} approaches unity (remember that $\rho_w \approx 1\,\text{g cm}^{-1}$), we derive the following rule:

$$\text{For ``poorly'' soluble compounds:} \quad C_w(\text{mol L}^{-1}) \approx 55.5X_s^s$$

Solubility Categories Aqueous solubilities are found to be expressed in categorical terms such as "practically insoluble", "slightly soluble", "soluble", "miscible", or similar terms. If a compound is miscible in any proportions, this is often denoted by "∞".

USES FOR SOLUBILITY DATA

- To estimate solubility in seawater (Section 11.8)
- To estimate air–water partition coefficients (Section 12.2)
- To estimate 1-octanol/water partition coefficients (Section 13.2)
- To estimate soil–water partition coefficients
- To estimate bioconcentration factors
- To estimate aquatic toxicology parameters
- To predict biodegradation potential of compounds [3]

Ionic Strength While most experimental solubility data have been determined in distilled, salt-free water, natural water usually contains various anionic and cationic species of mineral salts which change the electrolytic property of water and, hence, its capacity to dissolve organic compounds. Distilled water solubility and the solubility at different salt concentrations can be estimated knowing the ionic strength, I, of the solution. I is defined as follows:

$$I = \tfrac{1}{2}\sum(C_i Z_i^2) \tag{11.1.3}$$

where C_i and Z_i are the concentration and charge of the *i*th ionic species and the summation is over all ionic species present in the solution. Estimation of the seawater solubility from pure water solubility is presented in Section 11.8.

11.2 RELATIONSHIP BETWEEN ISOMERS

Aqueous solubilities depend strongly on the occurrence of functional groups with hydrogen-bonding capability, such as hydroxyl and amino groups. Isomers containing the same functional groups are expected to exhibit similar solubility behavior. For example, at 20°C 1-propanol and 2-propanol are both miscible with water at any proportions [4]. In Figures 11.2.1 to 11.2.4 further sets of isomers that are miscible

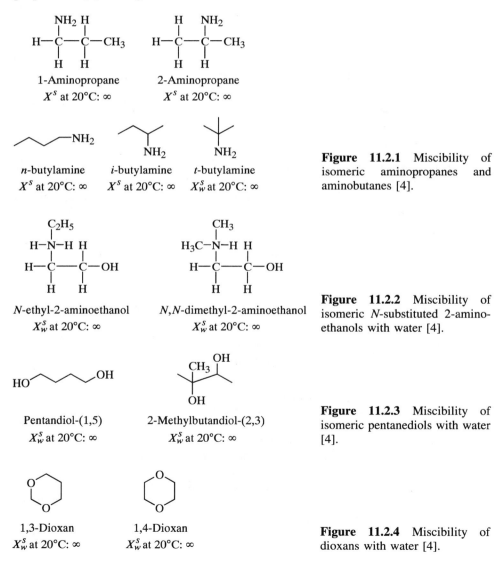

1-Aminopropane
X^s at 20°C: ∞

2-Aminopropane
X^s at 20°C: ∞

n-butylamine
X^s at 20°C: ∞

i-butylamine
X^s at 20°C: ∞

t-butylamine
X_w^s at 20°C: ∞

Figure 11.2.1 Miscibility of isomeric aminopropanes and aminobutanes [4].

N-ethyl-2-aminoethanol
X_w^s at 20°C: ∞

N,N-dimethyl-2-aminoethanol
X_w^s at 20°C: ∞

Figure 11.2.2 Miscibility of isomeric *N*-substituted 2-amino-ethanols with water [4].

Pentandiol-(1,5)
X_w^s at 20°C: ∞

2-Methylbutandiol-(2,3)
X_w^s at 20°C: ∞

Figure 11.2.3 Miscibility of isomeric pentanediols with water [4].

1,3-Dioxan
X_w^s at 20°C: ∞

1,4-Dioxan
X_w^s at 20°C: ∞

Figure 11.2.4 Miscibility of dioxans with water [4].

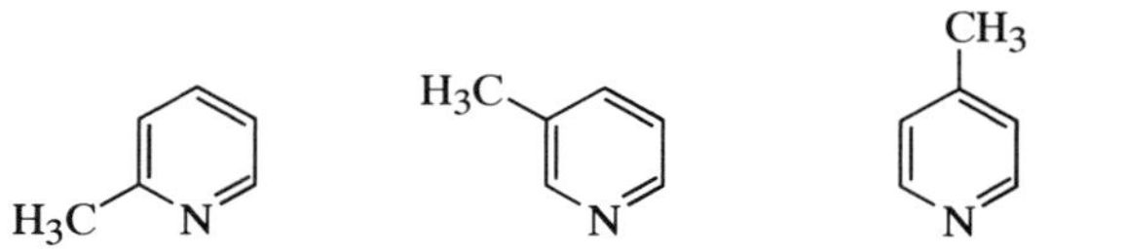

α-Picoline   β-Picoline   γ-Picoline
"soluble" at 20°C   miscible at 20°C   miscible at 20°C

Figure 11.2.5 Water solubility categories of picolines [4].

with water are shown. Figure 11.2.5 shows the solubility categories of picolines (i.e., methylpyridines). In α-picoline, where the methyl group is in close proximity to the nitrogen atom and partly inhibits its interaction with the water molecules, the water solubility is lowered. In contrast, the β and γ isomers are miscible with water in any proportions.

For nonmiscible compounds, the degree of branching and the position of functional groups in the molecule influences aqueous solubility of isomers. The following rules are representing selected examples:

Solubilities of branched alkanes are higher than the solubilitiy of their normal isomer. Solubilities increases with increasing degree of branching [5].   (R-11.2.1)

In linear, isomeric alkanols, the closer the OH group is to the center of the methylene chain, the more soluble is the alkanol [6].   (R-11.2.2)

The solubility of symmetrical n-alkyl n-alkoxypropionates is higher than the solubility of the isomeric n-alkyl methoxypropionate [7].   (R-11.2.3)

R-11.2.3 is illustrated in Figure 11.2.6 for ethyl ethoxypropionate having a higher solubility than either methyl n-propoxypropionate or n-propyl methoxypropionate.

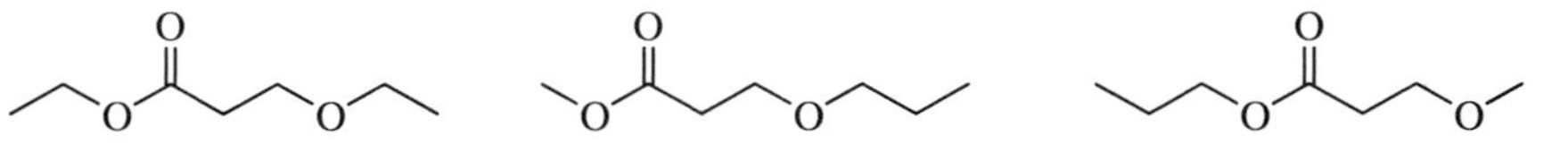

Ethyl ethoxypropionate   Methyl n-propoxypropionate   n-Propyl methoxypropionate
$S_w = 5.5$ $S_w = 3.4$ $S_w = 3.2$

Figure 11.2.6 Solubility (S_w in 100 mL of H_2O at room temperature) of isomeric n-alkyl β-n-alkoxypropionates [7].

11.3   HOMOLOGOUS SERIES AND AQUEOUS SOLUBILITY

A linear decrease of the aqueous solubility within several homologous series of hydrocarbons has been found. Coates et al. [8] reported the following correlations with C_w at 23°C:

for *n*-alkanes,

$$\log_{10}[C_w(\mu g\,mL^{-3})] = 4.416 - 0.569 N_C \qquad (11.3.1a)$$

for 2-methylalkanes,

$$\log_{10}[C_w(\mu g\,mL^{-3})] = 4.559 - 0.569 N_C \qquad (11.3.1b)$$

for 3-methylalkanes,

$$\log_{10}[C_w(\mu g\,mL^{-3})] = 4.376 - 0.559 N_C \qquad (11.3.1c)$$

and for 1-alkenes,

$$\log_{10}[C_w(\mu g\,mL^{-3})] = 5.108 - 0.569 N_C \qquad (11.3.1d)$$

Similar straight-line correlations between aqueous solubility and N_C or M have been found for certain homologous series of mono- and multifunctional compounds such as 1-alkanols, 2-alkanols [6], 2-alkanones [9], *n*-alkyl acetates [10], *n*-alkyl β-ethoxypropionates [11], *n*-alkyl α-acetoxypropionates, and *n*-alkyl lactates [12]. In contrast, Sobotka and Kahn [13] have found significant deviations from simple linear correlations with N_C for the series of ethyl esters of monocarbonic acids. They report a zig zag curve caused by the relatively higher solubility of the members with odd values for N_C. The odd–even effect is discussed in detail by Burrows [14], with further examples provided.

11.4   PROPERTY–SOLUBILITY RELATIONSHIPS

A diverse collection of quantitative property–water solubility relationships (QPWSR) is available in the literature. These QPWSR differ in their solubility representation (C_w, S_w, X_w), spectrum of independent variables, and applicability with respect to structure and physical state (liquid or solid). The following types of QPWSR are considered:

- Function of activity coefficient and crystallinity
- Solvatochromic approach
- Correlation with partition coefficient and melting point
- Correlation with boiling point
- Correlation with molar volume

Function of Activity Coefficients and Crystallinity   For compounds with very small water solubilities, the mole fraction solubility can be determined approximately by [15]

$$X_{w,s}^s \approx \frac{1}{\gamma_{w,s}^\infty} \qquad (11.4.1)$$

where $\gamma_{w,s}^{\infty}$ is the infinite dilution activity coefficient which may be calculated from the UNIFAC model.

A general model to estimate the mole fraction solubility of a solute s in water, $X_{w,s}^s$, is given by [16]

$$\log_{10} X_{w,s}^s = \frac{-\Delta S_f(T_m - T_x)}{2.303RT_x} - \log_{10}\gamma \qquad (11.4.2)$$

where ΔS_f is the solute's entropy of melting, T_m the melting point in K, R the universal gas constant, γ the activity coefficient, and T_x the temperature of interest in K. The second term on the right-hand side in eq. 11.4.2 is dependent on both solute and water properties, but the first term is solute specific and independent of water properties. Equation 10.4.2 applies over a broad compound range: organic non- and weak electrolytes and allows solubility estimation as a function of temperature. For the solubility of liquid or crystalline organic nonelectrolytes at 25°C, eq. 11.4.2 has been derived in a modified form [2]:

$$\log_{10}[C_w(\text{mol L}^{-1})] \approx 0.54 - 1.11\left[\frac{\Delta S_f(T_m - 25)}{1364}\right] - \log_{10}K_{ow} \qquad n = 167,$$
$$s = 0.242, \quad r = 0.994 \qquad (11.4.3)$$

where ΔS_f is in entropy units (eu), T_m is in °C, and K_{ow} is substituted for the activity coefficient. The middle term in eq. 10.4.2 diminishes for compounds with T_m equal or below 25°C. For rigid molecules a simpler model has been suggested that does not require the input of ΔS_f [2]:

$$\log_{10}[C_w(\text{mol L}^{-1})] = 0.87 - 0.012[T_m(°C)] - 1.05\log_{10}K_{ow} \qquad n = 155,$$
$$s = 0.308, \quad r = 0.989 \qquad (11.4.4)$$

A similar model has been reported for the solubility of mono- and polyhalogenated benzenes at 25°C [17]:

$$\log_{10}[C_w(\text{mol L}^{-1})] = 0.7178 - 0.0095[T_m(°C)] - 0.9874\log_{10}K_{ow} \qquad (11.4.5)$$

Replacing $\log_{10}K_{ow}$ by the total molecular surface area, TSA, the model is [17]

$$\log_{10}[C_w(\text{mol L}^{-1})] = 3.2970 - 0.0103[T_m(°C)] - 0.04225\text{TSA} \qquad (11.4.6)$$

A model to estimate solubilities for PCBs from T_m and TSA has been reported by Abramowitz and Yalkowsky [18]. This model is based on a method that allows T_m estimation from molecular structure input. Dunnivant et al. [19] have correlated T_m, TSA, and "third shadow area" with PCB solubility.

Molecular surface area is significant in relation to aqueous solubility and has been discussed by Amidon and Anik [20]. They have demonstrated the correlation of the molecular surface area with solution process parameters for hydrocarbons.

As illustrated with the model collection above, relatively simple models can be developed for hydrocarbons and certain classes of halogenated hydrocarbons, but the models for multi- and mixed-functional compounds require more involved parameter input.

Solvatochromic Approach   The solvatochromic approach describes a solvent-dependent property, XYZ, as a function of a cavity term, a dipolar term, and terms that account for hydrogen bonding [21]:

$$XYZ = XYZ_0 + \text{cavity term} + \text{dipolar term} + \text{hydrogen bonding term(s)}\quad(11.4.7)$$

where XYZ_0 is a compound-independent constant. The cavity term measures the free energy necessary to build a suitably sized cavity for a solute molecule between the solvent molecules. The dipolar term combines the solvatochromic parameters that measure solute–solvent, dipole–dipole, dipole–induced dipole, and dispersion interactions. The effect of hydrogen bonding is accounted for by a hydrogen-bond donor (HBD) and a hydrogen-bond acceptor (HBA) parameter, involving the solvent as donor and the solute as acceptor. Models based on the solvatochromic approach are frequently denoted as linear solvation energy relationships (LSERs).

LSER Model of Leahy   In the LSER model of Leahy [22], the cavity term is substituted by the molar volume, V_m, at 25°C in $g\,cm^{-3}$ or by the intrinsic molecular volume, V_i, in $mL\,mol^{-1}$. The dipolar term and the hydrogen-bonding terms are represented by the dipole moment, μ, and the HBA basicity, β, respectively. Group contribution schemes have been developed to calculate the solvatochromic parameters from molecular structure input [23]. Leahy [22] gives the following equation derived with a diverse set of monofunctional liquids:

$$\log_{10}[C_w(mol\,L^{-3})] = (0.23 \pm 0.10) - \frac{(5.80 \pm 0.14)V_i}{100} + (1.00 \pm 0.14)\pi^*$$
$$+ (4.89 \pm 0.14)\beta \qquad n = 114, \quad s = 0.23, \quad r^2 = 0.975,$$
$$F = 1459 \tag{11.4.8}$$

where $\pi^* = 0.023 + 0.233\mu$. Leahy derived similar models for solids and gases.

   The solvatochromic approach has been criticized by Yalkowsky et al. [24]. In particular, they claim π^* to be an insignificant parameter for the estimation of aqueous solubilities and they contend that models in which the solubility is correlated with K_{ow} and T_m (models 11.4.3 to 11.4.5, 11.4.10 and 11.4.11) are more versatile and have a firmer thermodynamic basis.

LSER of He, Wang, Han, Zhao, Zhang, and Zou   The LSER model of He et al. [25] has been derived with 28 phenylsulfonyl alkanoates. It includes T_m as an independent variable:

$$-\log_{10}[C_w(mol\,L^{-3})] = (1.061 \pm 0.715) + \frac{(3.773 \pm 0.123)V_i}{100} - (1.051 \pm 0.189)\pi^*$$
$$- (2.071 \pm 0.323)\beta + (0.014 \pm 0.001)T_m \qquad n = 28,$$
$$s = 0.115, \quad r = 0.991 \tag{11.4.9}$$

where C_w is at 25°C and T_m is in °C.

Solubility–Partition Coefficient Relationships A critical review on the applicability of empirically derived solubility–K_{ow} models has been given by Yalkowsky et al. [24], Isnard and Lambert [26], Lyman [1], and Müller and Klein [27]. Equations 10.4.3 to 10.4.5 are examples of solubility–K_{ow} models. Isnard and Lambert developed a model based on 300 structurally diverse compounds. The model equation for liquids ($T_m < 25°C$) is

$$\log_{10}[C_w(\text{mol m}^{-3})] = 4.17 - 1.38 \log_{10} K_{ow}$$
$$n = 300, \quad s = 0.466, \quad r = 0.965 \quad\quad (11.4.10)$$

and for solids ($T_m > 25°C$) the equation is

$$\log_{10}[C_w(\text{mol m}^{-3})] = 4.00 - 1.26 \log_{10} K_{ow} - 0.0054[T_m(°C) - 25°C]$$
$$n = 300, \quad s = 0.582, \quad r = 0.965 \quad\quad (11.4.11)$$

If these equations are applied in combination with structure-based methods to estimate K_{ow}, then only T_m or merely the information of liquidity is required as input to 11.4.11 or 11.4.10, respectively.

Solubility–Boiling Point Relationships Aqueous solubilities have been represented as polynomial functions of normal boiling points for alkanes and cycloalkanes. Yaws et al. [28] give the following equation:

$$\log_{10}[S_w(\text{ppmw})] = A + BT_b + CT_b^2 + DT_b^3 \quad\quad (11.4.12)$$

where T_b is in K. Coefficients A, B, C, and D are listed for different solubility temperatures in Table 11.4.1 for alkanes (C_5–C_{17}) and for alkyl-substituted cyclopentanes and cyclohexanes (C_5–C_{15}).

TABLE 11.4.1 Alkane and Cycloalkane Coefficients for eq. 11.4.12

T of $S_w(°C)$	A	B	C	D
		Alkanes		
25.0	−17.652	0.177811	−500.907·10^{-6}	411.124·10^{-9}
99.1	−17.261	0.177811	−500.907·10^{-6}	411.124·10^{-9}
121.3	−0.736	0.0411139	−136.980·10^{-6}	170.019·10^{-9}
		Cyclopentanes		
25.0	−16.900	177.88·10^{-3}	−500.907·10^{-6}	411.124·10^{-9}
99.1	−16.567	177.88·10^{-3}	−500.907·10^{-6}	411.124·10^{-9}
120.0	−0.033	−411.139·10^{-4}	−136.980·10^{-6}	170.019·10^{-9}
		Cyclohexanes		
25.0	−16.700	177.88·10^{-3}	−500.907·10^{-6}	411.124·10^{-9}
99.1	−16.290	177.88·10^{-3}	−500.907·10^{-6}	411.124·10^{-9}
120.0	−0.085	411.139·10^{-4}	−136.980·10^{-6}	170.019·10^{-9}

Source: Refs. [28–30].

Miller et al. [30] derived the following equation for chlorobenzenes at 25°C:

$$\log_{10}[C_w(\text{mol L}^{-1})] = 0.378 - 0.00211[T_b(°C)] \qquad n = 12, \quad r = 0.943$$

$$(11.4.13)$$

Almgren et al. [31] have reported a similar but more general correlation for aromatic hydrocarbons, including alkylbenzenes, chlorobenzenes, biphenyl, alkylnapthalenes, and PAHs up to five rings. The solubility is at 25°C:

$$\log_{10}[C_w(\text{mol L}^{-1})] = 0.76 - 0.0138[T_b(°C)] \qquad n = 29, \quad r = 0.97 \quad (11.4.14)$$

This model does not apply for molecules with a long aliphatic chain such as n-butylbenzene, polycyclic aromatic hydrocarbon compounds in which the rings are fused linearly, such as anthracene and chrysene.

Solubility–Molar Volume Relationships The correlation between aqueous solubility at room temperature and the molar volume has been studied by McAuliffe [5] for different hydrocarbon classes. He discusses linear relationships, presented as graphs, describing the decrease in solubility with increasing molar volume for the homologous series of alkanes, alkenes, alkandienes, alkynes, and cycloalkanes.

11.5 STRUCTURE–SOLUBILITY RELATIONSHIPS

The correlation between aqueous solubility and molar volume discussed by McAuliffe [5] for hydrocarbons, and the importance of the cavity term in the solvatochromic approach, indicates a significant solubility dependence on the molecular size and shape of solutes. Molecular size and shape parameters frequently used in quantitative structure–water solubility relationships (QSWSRs) are molecular volume and molecular connectivity indices. Moriguchi et al. [33] evaluated the following relationship to estimate C_w of apolar compounds and a variety of derivatives with hydrophilic groups:

$$\log_{10}[C_w(\text{mol L}^{-1})] = 0.98(\pm 0.14) - 3.95(\pm 0.19)V_L$$
$$n = 156, \quad s = 0.380 \quad r = 0.386 \qquad (11.5.1)$$

where $V_L = V_{\text{vdW}} - V_H$, in which V_{vdW} is the van der Waals volume in Å^3 and V_H is the hydrophilic effect volume in Å^3. V_H is zero for apolar molecules. Derivation of V_H and V_w is described in the source [32].

Bhatnagar et al. [34] have found a significant correlation between C_w and V_{vdW} for alkanols (C_4–C_9):

$$\log[C_w(\text{mol L}^{-1})] = 6.908 - 8.596V_{\text{vdW}} \qquad n = 48, \quad s = 0.464, \quad r = 0.974,$$
$$F_{1,46} = 859.94$$
$$(11.5.2)$$

Patil [35] reports the following correlation for chlorobenzenes and PCBs at 25°C:

$$\log_{10}[C_w(\text{mol L}^{-1})] = -0.122 - 0.907\,^1\chi^v - 0.0299(^1\chi^v)^2$$
$$n = 71, \quad s = 9.4, \quad r = 0.98 \tag{11.5.3}$$

Nirmalakhandan and Speece [36] introduced the polarizability factor, Φ, as an additional molecular descriptor. They derived the following model for halogenated alkanes and alkenes, alkylbenzenes, halobenzenes, and alkanols:

$$\log_{10}[S_w(\text{g/g\%})] = 2.209 + 1.653\,^0\chi - 1.312\,^0\chi^v + 1.00\Phi$$
$$n = 145, \quad s = 0.318, \quad r = 0.962 \tag{11.5.4}$$

where Φ is given by:

$$\Phi = -0.361N_H - 0.963N_{Cl} + 0.767N_= \tag{11.5.4a}$$

This model is based on S_w data spanning 5 log units. Nirmalakhandan and Speece [36,37] discuss the model's validity and robustness in detail. They performed a test using experimental S_w data for esters, ethers, and aldehydes that were not included in the training set. They noted reasonably good agreement between experimental and estimated data for the test set and indicated that eq. 11.5.4 is applicable to dialkyl ethers, alkanals, and alkyl alkanoates, but not for ketones, amines, PAHs, and PCBs. Nirmalakhandan and Speece [37] expanded the model above for the PAHs, PCBs, and PCDDs. However, their model has been criticized by Yalkowsky and Mishra for incorrect and omitted data [38]. The revised model is [38]

$$\log_{10}[C_w^s(\text{mol L}^{-1})] = 1.564 + 1.627(^0\chi) - 1.372(^0\chi^v) + 1.000\Phi'$$
$$n = 470, \quad s = 0.355, \quad r = 0.990 \tag{11.5.5}$$

Φ' in eq. 11.5.5 is calculated as

$$\Phi' = -0.361N_H - 2.620N_F - 0.936N_{Cl} + 1.474N_I + 0.636N_{NH_2} + 0.833N_{NH}$$
$$- 1.695N_{NO_2} - 0.767N_= - 1.24I_A + 1.014I_K - 3.332I_D \tag{11.5.5a}$$

where I_A is an indicator for alkanes and alkenes, I_K an indicator for ketones and aldehydes, and I_D an indicators for dibenzodioxins.

Amidon et al. [39] have correlated the aqueous solubility of 127 aliphatic hydrocarbons, alcohols, ethers, aldehydes, ketones, fatty acids, and esters with their total molecular surface area:

$$\log_{10}[C_w^s(\text{mol L}^{-1})] = 4.44 - 0.0168\text{TSA} \qquad n = 127, \quad s = 0.216, \quad r = 0.988 \tag{11.5.6}$$

where TSA were calculated by the method of Hermann [40], including a solvent (water) molecule radius of 1.5 Å.

Müller and Klein [27] have compared the predictive capabilities of model 11.5.3 with selected, linear C_w^s versus K_{ow} regression models such as model 11.4.9. Known models of the latter type have usually been derived from "mixed" K_{ow} data (i.e., K_{ow} is either estimated, experimental, or an average of several values, depending on what information is available for a compound). Müller and Klein derived a model for liquid compounds with unambiguous input:

$$\log_{10}[C_w(\text{mol L}^{-1})] = -1.16 - 0.79[\log_{10}(K_{ow})]_{\text{CLOGP}}$$
$$n = 156, \quad s = 0.298, \quad r = 0.950 \qquad (11.5.7)$$

where $[\log_{10}(K_{ow})]_{\text{CLOGP}}$ for all liquids is calculated solely from molecular structure input using the CLOGP algorithm. Comparing model 11.5.3, model 11.5.4, and five other C_w^s versus K_{ow} models by mean-square residual analysis using a validation set of over 300 liquid and solid compounds, they conclude that K_{ow}-based models, in general, yield more reliable results.

Nelson and Jurs [41] have developed models for three sets of compounds: (1) hydrocarbons, (2) halogenated hydrocarbons, and (3) alcohols and ethers. Each model correlates $\log[C_w^s(\text{mol L}^{-1})]$ with nine molecular descriptors that represent topological, geometrical, and electronic molecule properties. The standard error for the individual models is 0.17 log unit and for a fourth model that combines all three compound sets, the standard error is 0.37 log unit.

Bodor, et al. [42] compare the use of artificial neural networks with regression analysis techniques for the development of predictive solubility models. They report that the performance of the neural network model is superior to the regression-based model. Their study is based on a training set of 331 compounds. The model requires a diverse set of molecular descriptors to account for the structural variety in the training compounds.

11.6   GROUP CONTRIBUTION APPROACHES FOR AQUEOUS SOLUBILITY

Insertion of a methylene group into a molecule causes a decrease in aqueous solubility, however not with a universally applicable constant increment, as available GCMs might suggests. The odd–even effect (see Section 11.3) and the chain length have to be considered for accurate, quantitative estimations. In addition polar groups in the molecule affect the methylene contribution, as the following rule illustrates:

Insertion of a methylene group to an alkane, which is substituted with a polar group, decreases the aqueous solubility. The decrease   (R-11.6.1) depends on the polar group [6]: $-COOH > -NH_2 - OH$.

This effect is particularly pronounced between low N_C members of homologous compounds. Generally, this effect is regarded as secondary to group additivity.

The intramolecular group interaction in a solute molecule influences the aqueous solubility significantly. Henceforth, GCMs with a set of highly discriminative groups, which largely account for their structural group environment, would be desirable. The design of such GCMs is currently limited by the number of compounds that simultaneously contain specified groups and have measured data available. Thus GCM development has to seek a compromise between a precise, statistically robust model and a less precise model with a structurally broader applicability. This point was illustrated by Klopman et al. using GCMs for water solubility [40].

Methods of Klopman, Wang, and Balthasar   Klopman et al. [43] derived two GCMs for the estimation of S_w. Model I consist of 33 contribution parameters, whereas model II has 67 parameters. The equation for either model is

$$\log_{10}[S_w(\text{g/g\%})] = C_0 + \sum g_i G_i \qquad (11.6.1)$$

where C_0 is a constant, g_i the contribution coefficient from the ith group, and G_i the ith group, and the summation runs over all types i of contribution parameters. The values of the constants and the group contributions for models I and II are given in Appendix E. Both methods have been implemented in the Toolkit. Below the fragment constants are given for manual verification.

For 3-bromopropene, Figures 11.6.1 and 11.6.2 show the application of models I and II, respectively. An experimental C_w value of 3.17×10^{-2} mol L^{-1} at 25°C has been reported for 3-bromopropene [44]. For hexachlorobenzene, model I is illustrated in Figure 11.2.3. The following experimental C_w values have been found: $0.5 \times 10^{-5}, 3.5 \times 10^{-5}$, and 4.7×10^{-5} g L^{-1} [45].

Method of Wakita, Yoshimoto, Miyamoto, and Watanabe   The Wakita et al. method [46] has been derived with a set of 307 liquid compounds, including alkanes, alkenes, alkynes, halogenated alkanes, alkanols, oxoalkanes, alkanones, alkyl alkanoates, alkanethiols, alkanenitriles, nitroalkanes, and substituted benzenes, naphthalenes, and biphenyles. The model equation is

$$-\log_{10}[C_w(\text{mol L}^{-1})] = \sum g_i G_i \qquad (11.6.2)$$

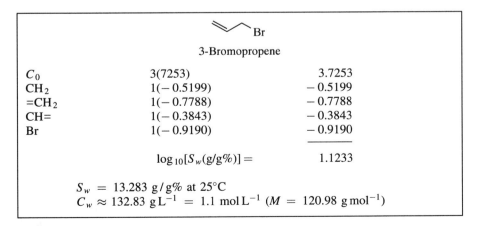

	3-Bromopropene	
C_0	3(7253)	3.7253
CH$_2$	1(−0.5199)	− 0.5199
=CH$_2$	1(−0.7788)	− 0.7788
CH=	1(−0.3843)	− 0.3843
Br	1(−0.9190)	− 0.9190
	$\log_{10}[S_w(\text{g/g\%})] =$	1.1233

$S_w = 13.283$ g/g% at 25°C
$C_w \approx 132.83$ g L^{-1} = 1.1 mol L^{-1} ($M = 120.98$ g mol^{-1})

Figure 11.6.1   Estimation of S_w (25°C) for 3-bromopropene using method I [43].

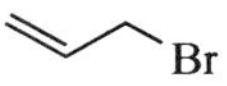

$$\nearrow\!\!\!\!\!\diagdown\text{Br}$$

3-Bromopropene

C_0	3.5650	3.5650
$-CH_2-$	$1(-0.5729)$	-0.5729
$=CH_2$	$1(-0.6870)$	-0.6870
$=CH-$	$1(-0.3230)$	-0.3230
$-Br$ (not connected to sp^3-C)	$1(-0.9643)$	-0.9643
	$\log_{10}[S_w(\text{g/g\%})] =$	1.0178

$S_w = 10.418$ g/g% at 25°C
$C_w \approx 104.18$ g L^{-1} = 0.861 mol L^{-1} ($M = 120.98$ g mol^{-1})

Figure 11.6.2 Estimation of S_w (25°C) for 3-bromopropene using method II [43].

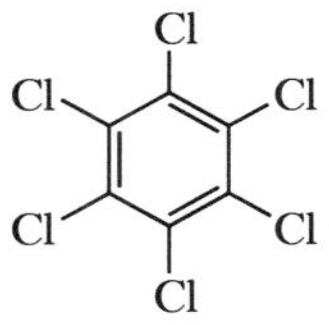

$$
\begin{array}{c}
\text{Cl} \\
\text{Cl} \diagdown\!\!\diagup \text{Cl} \\
\text{Cl} \diagup\!\!\diagdown \text{Cl} \\
\text{Cl}
\end{array}
$$

Hexachlorobenzene

C_0	3.5650	3.5650
$=C^*-(-)$	$6(-0.4944)$	-2.9664
$-Cl$ (not connected to sp^3-C)	$6(-0.6318)$	-3.7908
	$\log_{10}[S_w(\text{g/g\%})] =$	-3.1922

$S_w = 0.000642$ g/g% at 25°C
$C_w \approx 0.00642$ g L^{-1} = 2.25×10^{-5} mol L^{-1} ($M = 284.80$ g mol^{-1})

Figure 11.6.3 Estimation of S_w (25°C) for hexachlorobenzene using method I [43].

where the summation runs over all types i of contribution parameters. Contribution values have been evaluated in three steps: (1) aliphatic hydrocarbons, (2) substituted aliphatics, and (3) substituted aromatics. The contribution scheme is based on atom groups (aliphatic C, H, F, Cl, Br, I), functional groups (e.g., C=C, C≡C, C≡N, NO$_2$), ring contributions, and aromatic ring substituents (e.g., NH$_2$ in aniline derivatives). The correlation of the observed with the retro-estimated values is given by the following equation:

$$-\log_{10}[C_w(\text{mol L}^{-1})] = 0.957(\pm 0.021) \sum_i g_i G_i + 0.048(\pm 0.049)$$

$$n = 307, \quad s = 0.245, \quad r = 0.982 \tag{11.6.2a}$$

This method has been integrated into CHEMICALC2 [47] for automatic C_w and K_{ow} estimation (see the method of Suzuki and Kudo in Chapter 10).

AQUAFAC Approach   The AQUAFAC approach is based on the following solubility equation [48,49]:

$$\log C_{w,\text{obs}} = \log C_{w,\text{ideal}} + \log \gamma_w \qquad (11.6.3)$$

The ideal solubility $C_{w,\text{ideal}}$ in eq. 11.6.3 is expressed by

$$\log C_{w,\text{ideal}} = \Delta S_m (2.303 \times 298R)^{-1}(T_m - 25) \qquad (11.6.4)$$

where ΔS_m is the entropy of melting, T_m the melting point, and R the universal gas constant. The aqueous activity coefficient is a function of group contributions:

$$\log \gamma_w = \sum n_i q_i \qquad (11.6.5)$$

where q_i is the group contribution of type i and n_i is the number of times that group i appears in the molecule. Values for q have been derived for hydrocarbon, halogen, and non-hydrogen-bond-donating oxygen groups. The 27 group values have been derived from a set of 621 compounds representing over 1700 individual solubility values ranging from 3.60 to 3.47×10^{-13} mol L^{-1}. The overall statistics for this model are $n = 621$, $r^2 = 0.98$, RMSE $= 0.47$, $F = 1523$ [45]. Observed and calculated values for PCBs, chlorinated dibenzo-*p*-dioxins, and selected pesticides have been compared.

11.7   TEMPERATURE DEPENDENCE OF AQUEOUS SOLUBILITY

Aqueous solubility either increases or decreases with increasing temperature, depending on the considered temperature interval and the type of compounds. The temperature-dependence of the mole fraction aqueous solubility, X_s, for the equilibrium between organic phase and aqueous solution may be expressed by the van't Hoff equation:

$$\ln X_s = \frac{-\Delta H_{\text{soln}}}{RT} + C \qquad (11.7.1)$$

where ΔH_{soln} is the enthalpy of solution, R the gas constant, T the absolute temperature, and C is a constant. This equation applies to solutes below their melting point and for fairly small temperature ranges over which ΔH_{soln} remains relatively constant. Equation 11.7.1 is not valid when the water content in the organic phase changes with temperature. Dickhut et al. [50], for example, determined ΔH_{soln} and C for biphenyl, 4-chlorobiphenyl, and PCBs. Friesen and Webster [57] discusses the application of eq. 11.7.1 for polychlorinated dibenzo-*p*-dioxins between 7 and 41°C.

Wauchope and Getzen [52] employ a semiempirical function including the molar heat of fusion to fit $X_w = f(T)$ for PAHs. May et al. [53] employ an empirical, cubic temperature function for PAHs. A quadratic function has been derived by Yaws et al. [29] for alkanes (C_5–C_{17}) and cycloalkanes (C_5–C_{15}):

$$\log_{10}[S_w(\text{ppmw})] = A + BT^{-1} + CT^{-2} \qquad (11.7.2)$$

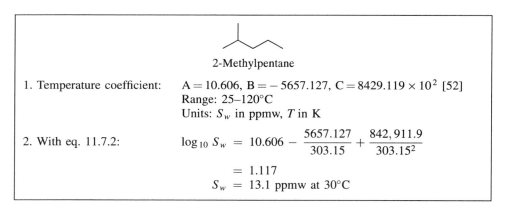

2-Methylpentane

1. Temperature coefficient: $A = 10.606$, $B = -5657.127$, $C = 8429.119 \times 10^2$ [52]
Range: 25–120°C
Units: S_w in ppmw, T in K

2. With eq. 11.7.2: $\log_{10} S_w = 10.606 - \dfrac{5657.127}{303.15} + \dfrac{842,911.9}{303.15^2}$

$= 1.117$

$S_w = 13.1$ ppmw at 30°C

Figure 11.7.1   Estimation of S_w (30°C) for 2-methylpentane.

where T is in K. This equation applies in the temperature range from 25 to 120°C. The coefficients of eq. 11.7.2 are included in the Toolkit permitting the calculation of S_w at the specified temperature. The estimation of S_w at 30°C is illustrated in Figure 11.7.1. Howe et al. [54] reported an experimental value of 16 ppmw.

For biphenyls, dibenzofurans, dibenzo-p-dioxins, and their halogenated derivatives, Doucette and Andren [55] fitted solubility data with the following equation:

$$C_w (\text{mol L}^{-1}) = a + e^{b[T(°C)]} \tag{11.7.3}$$

where a and b are compound-specific constants. This equation is incorporated into the Toolkit to estimate C_w at specified temperature in the range 4 to 40°C for the corresponding compounds. For the same sets of compounds, Doucette and Andren derived an equation that allows the temperature-dependent estimation of C_w independent from any compound-specific parameters besides C_w at the reference temperature 25°C:

$$\log_{10}[C_w^T (\text{mol L}^{-1})] = \log_{10}[C_w^{25} (\text{mol L}^{-1})] + 0.0235[T(°C)] - 0.588 \tag{11.7.4}$$

Estimation from Henry's Law Constant   In certain cases the water solubility at temperature T can be calculated as the ratio of the liquid vapor pressure at saturation, P_L^s, and H_c, or as the ratio of the solid vapor pressure at saturation, P_s^s, and H_c, if these data are known at T:

For solid compounds: $$C_{w,s} = \frac{P_s^s}{H_c} \tag{11.7.5a}$$

For liquid compounds: $$C_{w,L} = \frac{P_L^s}{H_c} \tag{11.7.5b}$$

This approach has been applied for hydrocarbons, halogenated hydrocarbons, and various classes of pesticides. However, if the solute and water are mutually soluble into each other in appreciable amounts (e.g., > 5% mol), these equations are no longer

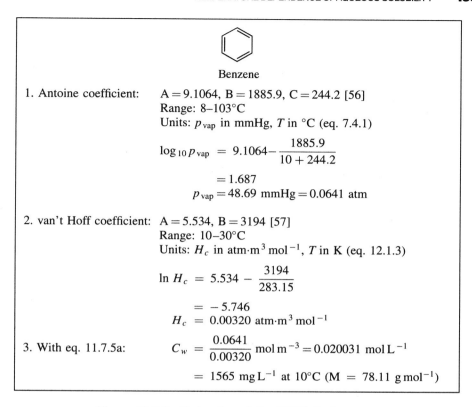

Benzene

1. Antoine coefficient: A = 9.1064, B = 1885.9, C = 244.2 [56]
Range: 8–103°C
Units: p_{vap} in mmHg, T in °C (eq. 7.4.1)

$$\log_{10} p_{vap} = 9.1064 - \frac{1885.9}{10 + 244.2}$$

$$= 1.687$$
$$p_{vap} = 48.69 \text{ mmHg} = 0.0641 \text{ atm}$$

2. van't Hoff coefficient: A = 5.534, B = 3194 [57]
Range: 10–30°C
Units: H_c in atm·m^3 mol^{-1}, T in K (eq. 12.1.3)

$$\ln H_c = 5.534 - \frac{3194}{283.15}$$

$$= -5.746$$
$$H_c = 0.00320 \text{ atm·m}^3 \text{ mol}^{-1}$$

3. With eq. 11.7.5a: $$C_w = \frac{0.0641}{0.00320} \text{ mol m}^{-3} = 0.020031 \text{ mol L}^{-1}$$

$$= 1565 \text{ mg L}^{-1} \text{ at } 10°C \text{ (M} = 78.11 \text{ g mol}^{-1}\text{)}$$

Figure 11.7.2   Estimation of C_w (10°C) for benzene.

valid. The Toolkit utilizes the temperature functions of vapor pressure and air–water partition coefficients and applies eqs. 11.7.5a and 11.7.5b to estimate C_w. An example is given for benzene in Figure 11.7.2. An experimental value of 1822 ppmw at 10°C has been reported [54].

Compounds with a Minimum in Their S(T) function   Many compounds, such as alkanes and its derivatives with a hydrophilic polar group, exhibit a solubility minimum at a temperature T_{min}. For liquid alkanols, alkanoic acids, and alkylamines, T_{min} lies between 15 and 80°C [6]. The following qualitative results were given:

For each of the series of liquid alkanols, alkanoic acids, and alkyl-amines, T_{min} decreases with increasing N_C [6].   (R-11.7.1)

For constant N_C ($N_C = 5$ or $N_C = 6$) the following order applies [6]:

$$T_{min}(\text{3-alkanol}) > T_{min}(\text{2-alkanol}) > T_{min}(\text{1-aminoalkane})$$ (R-11.7.2)
$$> T_{min}(\text{1-alkanol}) > T_{min}(\text{alkanoic acid})$$

$T_{\min}$ can be estimated quantitatively, if the enthalpy of evaporation, ΔH_v, is known. For alkanols, alkanoic acids, and alkylamines, ΔH_v is constant in the temperature range between 305 and 445°C and the following relation has been derived:

$$T_{\min}(\text{K}) = \frac{233}{\Delta H_{\text{vap}} - 41.0} + 288 \tag{11.7.6}$$

where ΔH_{vap} is in kJ mol^{-1}. This equation applies for compounds having ΔH_v values higher than 41.0 kJ mol^{-1} and $T_{\min}$ values greater than 288 K [6].

The estimation of the aqueous solubility at $T_{\min}$ and at other temperatures requires data on the enthalpy and the heat capacity of the solution. These properties are themselves temperature dependent and have been systematically studied for various sets of compounds such as hydrocarbons [58,59], 1-alkanols [60], alkoxyethanols, and 1,2-dialkoxyethanes [61], carboxylic acids, amines, and N-substituted amides [62], monoesters, ethylene glycol diesters, glycerol triesters [63], and crown ethers [64]. Additive schemes for the estimation of aqueous solution heat capacities have been evaluated [65,66].

Quantitative Property–S_w(T) Relationship Dickhut et al. [67] developed a QP–$S_w(T)$R based on experimental mole fraction solubilities for alkylbenzenes, PAHs, PCBs, chlorinated dibenzofuranes and p-dioxins, and alkyl- and halosubstituted naphthalenes and p-terphenyls in the range 4 to 40°C:

$$\ln x_l = -0.0807\text{TSA} \tag{11.7.7a}$$

$$\ln x_g = -\left[6.5922\left(\frac{T_m}{T} - 1\right)\right] - 0.0807\text{TSA} \tag{11.7.7b}$$

where x_l and x_g are the mole fraction solubilities for liquids and solids, respectively; TSA is the total surface area; T_m is the melting point in K; and T is the temperature of interest in K.

11.8 SOLUBILITY IN SEAWATER

Seawater contains dissolved inorganic salts. An aqueous solution of about 35 gL^{-1} NaCl is often taken as a model solution for seawater. The salt effect on the solubility of nonelectrolyte organic compounds has been investigated systematically by Sechenov [68] and by Long and McDevit [69]. Correlations between pure water solubility, S_w, and the solubility at different salt concentrations are compound dependent. For example, the seawater solubility, S_{sw}, of PAHs are from 30 to 60% below their freshwater solubilities [1], depending on the particular structure of the PAH. We concentrate our interest on the question if, for certain compound classes, S_{sw} can be estimated from known S_w without any input of further compound-specific parameters.

Sutton and Calder [70] measured S_w and S_{sw} for several n-alkanes and alkyl benzenes at 25°C and reported that in all cases $S_{\text{sw}} < S_w$. Similarly, Groves [71] found that the salt water solubility (34.5 parts of NaCl per thousand parts of water) at 25°C of cyclopentane, cyclohexane, methylcyclohexane, and cycloheptane is lower than S_w. Keeley et al. [72] studied the solubility of benzene and toluene in aqueous NaCl solution at 25°C in the ionic strength range 0 to 5. For both compounds the solubility decreases with increasing ionic strength. The same trend is found for hexane, phenanthrene, chlorobenzene, and 1,4-dichlorobenzene in solutions of NaCl, KCl, NH$_4$Cl, NaBr, and Na$_2$SO$_4$ [73]. We summarize the foregoing observations in the following rule:

A salting-out effect occurs for alkanes, cycloalkanes, alkylbenzenes, PAHs, and chlorobenzenes: $S_{sw} < S_w$ (R-11.8.1)

A quantitative correlation has been derived based on experimental data for 11 aromatic compounds (biphenyl, naphthalene, anthracene, phenanthrene, pyrene, phenol, *p*-toluidine, *p*-nitrotoluene, and *o*-, *m*-, and *p*-nitrophenol) at 20°C [74]:

$$\log_{10}[S_{sw}(\text{mol L}^{-1})] = (0.0298I + 1)\log_{10}[S_w(\text{mol L}^{-1})] - 0.114I \qquad (11.8.1)$$

where I is the ionic strength in mol L^{-1} for the solution of interest. Equation 11.8.1 has been suggested for the estimation of S_{sw} in the range 10^{-7} to 1 mol L^{-1}.

REFERENCES

1. Lyman, W. J., Solubility in Water, in *Handbook of Chemical Property Estimation Methods*, W. J. Lyman, W. F. Reehl, and D. H. Rosenblatt, Editors, 1990. Washington, DC: American Chemical Society.

2. Yalkowsky, S. H., and S. C. Valvani, Solubility and Partitioning: I. Solubility of Noneelctrolytes in Water. *J. Pharm. Sci.*, 1980: **69**, 912–922.

3. Klopman, G., MULTICASE: 1. A Hierarchical Computer Automated Structure Evaluation Program. *Quant. Struct.-Act. Relat.*, 1992: **11**, 176–184.

4. Freier, R. K., *Aqueous Solutions*, Vol. 1, 1976. Berlin: Walter de Gruyter.

5. McAuliffe, C., Solubility in Water of Paraffin, Cycloparaffin, Olefin, Acetylene, Cyclo-olefin, and Aromatic Hydrocarbons. *J. Phys. Chem.*, 1966: **70**, 1267–1275.

6. Nishino, N., and M. Nakamura, Minima in the Solubilities of Normal Alkane Derivatives with a Polar Group in Water. *Bull. Chem. Soc. Jpn.*, 1978: **51**, 1617–1620.

7. Rehberg, C. E., M. B. Dixon, and C. H. Fisher, Preparation and Physical Properties of *n*-Alkyl-*β*-*n*-Alkoxypropionates. *J. Am. Chem. Soc.*, 1947: **69**, 2966–2970.

8. Coates, M., D. W. Connell, and D. M. Barron, Aqueous Solubility and Octan-1-ol to Water Partition Coefficients of Aliphatic Hydrocarbons. *Environ. Sci. Technol.*, 1985: **19**, 628–632.

9. Erichsen, L. V., Die Löslichkeit homologer Reihen organischer Verbindungen. *Naturwissenschaften*, 1952: **39**, 41–42.

10. Erichsen, L. V., Die Löslichkeit homologer Reihen organischer Verbindungen: 2. Das Löslichkeitsdekrement der Methylengruppe und die Funktionslöslichkeit in homologen Reihen. *Naturwissenschaften*, 1952: **39**, 189.

11. Dixon, M. B., C. E. Rehberg, and C. H. Fisher, Preparation and Physical Properties of *n*-Alkyl-*β*-Ethoxypropionates. *J. Am. Chem. Soc.*, 1948: **70**, 3733–3738.

12. Rehberg, C. E., and M. B. Dixon, *n*-Alkyl Lactates and their Acetates. *J. Am. Chem. Soc.*, 1950: **72**, 1918–1922.

13. Sobotka, H., and J. Kahn, Determination of Solubility of Sparingly Soluble Liquids in Water. *J. Am. Chem. Soc.*, 1931: **53**, 2935–2938.

14. Burrows, H. D., Studying Odd–Even Effects and Solubility Behavior Using α,ω-Dicarboxylic Acids. *J. Chem. Educ.*, 1992: **69**, 69–73.

15. Chen, F., J. Holten-Andersen, and H. Tyle, New Developments of the UNIFAC Model for Environmental Application. *Chemosphere*, 1993: **26**, 1325–1354.

16. Yalkowsky, S. H., Estimation of the Aqueous Solubility of Complex Organic Compounds. *Chemosphere*, 1993: **26**, 1239–1261.

17. Yalkowsky, S. H., R. J. Orr, and S. C. Valvani, Solubility and Partitioning: 3. The Solubility of Halobenzenes in Water. *Ind. Eng. Chem. Fundam.*, 1979: **18**, 351–353.

18. Abramowitz, R., and S. H. Yalkowsky, Estimation of Aqueous Solubility and Melting Point of PCB Congeners. *Chemosphere*, 1990: **21**, 1221–1229.

19. Dunnivant, F. M., et al., Quantitative Structure–Property Relationships for Aqueous Solubilities and Henry's Law Constants of Polychlorinated Biphenyls. *Environ. Sci. Technol.*, 1992: **26**, 1567–1573.

20. Amidon, G. L., and S. T. Anik, Application of the Surface Area Approach to the Correlation and Estimation of Aqueous Solubility and Vapor Pressure: Alkyl Aromatic Hydrocarbons. *J. Chem. Eng. Data*, 1981: **26**, 28–33.

21. Kamlet, M. J., et al., Linear Solvation Energy Relationships: 36. Molecular Properties Governing Solubilities of Organic Nonelectrolytes in Water. *J. Pharm. Sci.*, 1986: **75**, 338–349.

22. Leahy, D. E., Intrinsic Molecular Volume as a Measure of the Cavity Term in Linear Solvation Energy Relationships: Octanol–Water Partition Coefficients and Aqueous Solubilities. *J. Pharm. Sci.*, 1986: **75**, 629–636.

23. Hickey, J. P., and D. R. Passino-Reader, Linear Solvation Energy Relationships: "Rules of Thumb" for Estimation of Variable Values. *Environ. Sci. Technol.*, 1991: **25**, 1753–1760.

24. Yalkowsky, S. H., R. Pinal, and S. Banerjee, Water Solubility: A Critique of the Solvatochromic Approach. J. Pharm. Sci., 1988: **77**, 74–77.

25. He, Y., et al., Determination and Estimation of Physicochemical Properties for Phenylsulfonyl Acetates. *Chemosphere*, 1995: **30**, 117–125.

26. Isnard, P., and S. Lambert, Aqueous Solubility and *n*-Octanol/Water Partition Coefficient Correlations. *Chemosphere*, 1989: **18**, 1837–1853.

27. Müller, M., and W. Klein, Comparative Evaluation of Methods Predicting Water Solubility for Organic Compounds. *Chemosphere*, 1992: **25**, 769–782.

28. Yaws, C. L., et al., Hydrocarbons: Water Solubility Data. *Chem. Eng.*, 1990: Apr., 177–181.

29. Yaws, C. L., et al., Water Solubility Data for 151 Hydrocarbons. *Chem. Eng.*, 1993: Feb., 108–111.

30. Yaws, C. L., X. Lin, and L. Bu, The Water Solubility of Naphthenes. *Chem. Eng.*, 1993: Oct., 122–123.

31. Miller, M. M., et al., Aqueous Solubilities, Octanol/Water Partition Coefficients, and Entropies of Melting of Chlorinated Benzenes and Biphenyls. *J. Chem. Eng. Data*, 1984: **29**, 184–190.

32. Almgren, M., et al., A Correlation Between the Solubility of Aromatic Hydrocarbons in Water and Micellar Solutions, with Their Normal Boiling Points. *J. Chem. Eng. Data*, 1979: **24**, 285–287.

33. Moriguchi, I., Y. Kanada, and K. Komatsu, Van der Waals Volume and the Related Parameters for Hydrophobicity in Structure-Activity Studies. *Chem. Pharm. Bull.*, 1976: **24**, 1799–1806.

34. Bhatnagar, R. P., P. Singh, and S. P. Gupta, Correlation of van der Waals Volume with Boiling Point, Solubility and Molar Refraction. *Indian J. Chem.*, 1980: **19B**, 780–783.

35. Patil, G. S., Correlation of Aqueous Solubility and Octanol-Water Position Coefficient Based on Molecular Structure. *Chemosphere*, 1991: **22**, 723–738.

36. Nirmalakhandan, N. N., and R. E. Speece, Prediction of Aqueous Solubility of Organic Chemicals Based on Molecular Structure. *Environ. Sci. Technol.*, 1988: **22**, 328–338.

37. Nirmalakhandan, N. N., and R. E. Speece, Prediction of Aqueous Solubility of Organic Chemicals Based on Molecular Structure. 2. Application to PNAs, PCBs, PCDDs, etc. *Environ. Sci. Technol.* 1989: **23**, 708–713.

38. Speece, R. E., Comment on "Prediction of Aqueous Solubility of Organic Chemicals Based on Molecular Structure. 2. Application to PNAs, PCBs, PCDDs, etc." *Environ. Sci. Technol.*, 1990: **24**, 927–929.

39. Amidon, G. L., et al., Solubility of Nonelectrolytes in Polar Solvents: V. Estimation of the Solubility of Aliphatic Monofunctional Compounds in Water Using a Molecular Surface Area Approach. *J. Phys. Chem.*, 1975: **79**, 2239–2246.

40. Hermann, R. B., Theory of Hydrophobic Bonding: II. The Correlation of Hydrocarbon Solubility in Water with Solvent Cavity Surface Area. *J. Phys. Chem.*, 1972: **75**, 2754–2759.

41. Nelson, T. M., and P. C. Jurs, Prediction of Aqueous Solubility of Organic Compounds. *J. Chem. Inf. Comput. Sci.*, 1994: **34**, 601–609.

42. Bodor, N., A. Harget, and M.-J. Huang, Neural Network Studies: 1. Estimation of the Aqueous Solubility of Organic Compounds. *J. Am. Chem. Soc.*, 1991: **113**, 9480–9483.

43. Klopman, G., S. Wang, and D. M. Balthasar, Estimation of Aqueous Solubility of Organic Molecules by the Group Contribution Approach: Application to the Study of Biodegradation. *J. Chem. Inf. Comput. Sci.*, 1992: **32**, 474–482.

44. Tewari, Y. B., et al., Aqueous Solubility and Octanol/Water Partition Coefficient of Organic Compounds at 25.0°C. *J. Chem. Eng. Data*, 1982: **27**, 451–454.

45. Suntio, L. R., et al., Critical Review of Henry's Law Constants for Pesticides. *Rev. Environ. Contam. Toxicol.*, 1988: **103**, 1–59.

46. Wakita, K., et al., A Method for Calculation of the Aqueous Solubility of Organic Compounds by Using New Fragment Solubility Constants. *Chem. Pharm. Bull.*, 1986: **34**, 4663–4681.

47. Suzuki, T., Development of an Automatic Estimation System for Both the Partition Coefficient and Aqueous Solubility. *J. Comput.-Aid. Mol. Des.*, 1991: **5**, 149–166.

48. Myrdal, P., et al., AQUAFAC: Aqueous Functional Group Activity Coefficients. *SAR QSAR Environ. Res.*, 1993: **1**, 53–61.

49. Myrdal, P., and S. H. Yalkowsky, A Simple Scheme for Calculating Aqueous Solubility, Vapor Pressure and Henry's Law Constant: Application to the Chlorobenzenes. *SAR QSAR Environ. Res.*, 1994: **2**, 17–28.

50. Dickhut, R. M., A. W. Andren, and D.E. Armstrong, Aqueous Solubilities of Six Polychlorinated Biphenyl Congeners at Four Temperatures. *Environ. Sci. Technol.*, 1986: **20**, 807–810.

51. Friesen, K. J., and G. R. B. Webster, Temperature Dependence of the Aqueous Solubilities of Highly Chlorinated Dibenzo-*p*- Dioxins. *Environ. Sci. Technol.*, 1990: **24**, 97–101.

52. Wauchope, R. D., and F. W. Getzen, Temperature Dependence of Solubilities in Water and Heats of Fusion of Solid Aromatic Hydrocarbons. *J. Chem. Eng. Data*, 1972: **17**, 38–41.

53. May, W. E., S. P. Wasik, and D. H. Freeman, Determination of the Aqueous Solubility of Polynuclear Aromatic Hydrocarbons by a Coupled Column Liquid Chromatographic Technique. *Anal. Chem.*, 1978: **50**, 175–179.

54. Howe, G. B., M. E. Mullins, and T. N. Rogers, *Evaluation and Prediction of Henry's Law Constants and Aqueous Solubilities for Solvents and Hydrocarbon Fuel Components*: Vol. I. *Technical Discussion*, Report ESL-TR-86-66, 1987. Research Triangle Park, NC: Research Triangle Institute. (Request copies from National Technical Information Service, 5285 Port Royal Road, Springfield, VA 22161.)

55. Doucette, W. J., and A. W. Andren, Aqueous Solubility of Selected Biphenyl, Furan, and Dioxin Congeners. *Chemosphere*, 1988: **17**, 243–252.

56. Dean, J. A., *Lange's Handbook of Chemistry*, 14th ed., 1992. New York: McGraw-Hill.

57. Ashworth, R. A., et al., Air–Water Partitioning Coefficients of Organics in Dilute Aqueous Solutions. J. Hazard. Mater., 1988: **18**, 25–36.

58. Owens, J. W., S. P. Wasik, and H. DeVoe, Aqueous Solubilities and Enthalpies of Solution of *n*-Alkylbenzenes. *J. Chem. Eng. Data*, 1986: **31**, 47–51.

59. Gill, S. J., N. F. Nichols, and I. Wadsö, Calorimetric Determination of Enthalpies of Solution of Slightly Soluble Liquids: II. Enthalpy of Solution of Some Hydrocarbons in Water and Their Use in Establishing the Temperature Dependence of Their Solubilities. *J. Chem. Thermodyn.*, 1976: **8**, 445–452.

60. Hallen, D., et al., Enthalpies and Heat Capacities for *n*-Alkan-1-ols in H_2O and D_2O. *J. Chem. Thermodyn.*, 1986: **18**, 429–442.

61. Kusano, K., J. Suurkuusk, and I. Wadsö, Thermochemistry of Solutions of Biochemical Model Compounds: 2. Alkoxyethanols and 1,2- Dialkoxyethanes in Water. *J. Chem. Thermodyn.*, 1973: **5**, 757–767.

62. Konicek, J., and I. Wadsö, Thermochemical Properties of Some Carboxylic Acids, Amines and N-Substituted Amides in Aqueous Solution. *Acta Chim. Scand.*, 1971: **25**, 1541–1551.

63. Nilsson, S.-O. and I. Wadsö, Thermodynamic Properties of Some Mono-, Di-, and Triesters: Enthalpies of Solution in Water at 288.15 to 318.15 K and Enthalpies of Vaporization and Heat Capacities at 298.15 K. *J. Chem. Thermodyn.*, 1986: **18**, 673–681.

64. Briggner, L.-E. and I. Wadsö, Some Thermodynamics Properties of Crown Ethers in Aqueous Solution. *J. Chem. Thermodyn.*, 1990: **22**, 143–148.

65. Nichols, N., et al., Additivity Relations for the Heat Capacities of Non-electrolytes in Aqueous Solution. *J. Chem. Thermodyn.*, 1976: **8**, 1081–1093.

66. Cabani, S., et al., Group Contribution to the Thermodynamic Properties of Non-ionic Organic Solutes in Dilute Aqueous Solution. *J. Solution Chem.*, 1981: **10**, 563–595.

67. Dickhut, R. M., K. E. Miller, and A. W. Andren, Evaluation of Total Molecular Surface Area for Predicting Air–Water Partitioning Properties of Hydrophobic Aromatic Chemicals. *Chemosphere*, 1994: **29**, 283–297.

68. Clever, H. L., The Ion Product Constant of Water: Thermodynamics of Water Ionization. *J. Chem. Educ.*, 1968: **45**, 231–235.

69. Long, F. A., and W. F. McDevit, Activity Coefficients of Nonelectrolyte Solutes in Aqueous Salt Solutions. *Chem. Rev.*, 1952: **51**, 119–169.

70. Sutton, C., and J. A. Calder, Solubility of High-Molecular-Weight *n*-Paraffins in Distilled Water and Seawater. *Environ. Sci. Technol.*, 1974: **8**, 655–657.

71. Groves, F. R., and Jr., Solubility of Cycloparaffins in Distilled Water and Salt Water. *J. Chem. Eng. Data*, 1988: **33**, 136–138.

72. Keeley, D. F., M. A. Hoffpauir, and J. R. Meriwether, Solubility of Aromatic Hydrocarbons in Water and Sodium Chloride Solutions of Different Ionic Strength: Benzene and Toluene. *J. Chem. Eng. Data*, 1988: **33**, 87–89.

73. Aquan-Yuen, M., D. Mackay, and W. Y. Shiu, Solubility of Hexane, Phenanthrene, Chlorobenzene, and *p*-Dichlorobenzene in Aqueous Electrolyte Solutions. *J. Chem. Eng. Data*, 1979: **24**, 30–34.

74. Hashimoto, Y., et al., Prediction of Seawater Solubility of Aromatic Compounds. *Chemosphere*, 1984: **13**, 881–888.

CHAPTER 12

AIR–WATER PARTITION COEFFICIENT

12.1 DEFINITIONS

The equilibrium air–water partition coefficient (AWPC), can be defined in different forms. Frequently used is the concentration to concentration ratio, K_{aw}:

$$K_{aw} = \frac{C_a}{C_w} \tag{12.1.1}$$

where C_a is the vapor-phase concentration, C_w is the aqueous-phase concentration and C_a and C_w are in $mol\,L^{-1}$, $\mu g\,L^{-1}$, or equivalent units. Therefore, K_{aw} is dimensionless. Alternatively, the AWPC can be expressed as the partial pressure to liquid concentration ratio, the Henry's law constant, H_c:

$$H_c = \frac{p_i}{C_w} \tag{12.1.2}$$

where p_i is the compound's gas-phase pressure in units of atm or kPa and C_w is in $mol\,L^{-1}$. H_c is related to K_{aw} through the ideal gas law:

$$H_c = K_{aw}RT \tag{12.1.3}$$

where T is the absolute temperature in K and R is the gas constant, equal to $8.314\,kPa{\cdot}m^3\,mol^{-1}\,K^{-1}$ or to $82.06 \times 10^{-6}\,atm{\cdot}m^3\,mol^{-1}\,K^{-1}$, depending on whether H_c is in $kPa{\cdot}m^3\,mol^{-1}$ or in $atm{\cdot}m^3\,mol^{-1}$, respectively. Two other expressions for the AWPC, H_y, and H_x are in use [1,2]:

$$H_y(-) = \frac{y_i}{x_i} \tag{12.1.4}$$

$$H_x(atm) = \frac{y_i}{x_i}p_G(atm) \tag{12.1.5}$$

where y_i and x_i are the gas- and liquid-phase mole fraction, respectively, and p_G is the total (atmospheric) gas-phase pressure. H_y is related to K_{aw} by the following equation [1]:

$$H_y = K_{aw} \frac{RT}{p_G V_s} \qquad (12.1.6)$$

where R is 82.06 atm·m^3 mol^{-1} K^{-1}, T is in K, p_G is the pressure in atm, and V_s is the molar volume of the solution in m^3 mol^{-1}.

The experimental techniques available to determine AWPCs and their limitations have been discussed by Staudinger and Roberts [2]. These authors also evaluated the effects of pH, compound hydration, compound concentration, cosolvent, cosolute, and salt effects, suspended solids, dissolved organic matter, and surfactants. The experimental data have been compiled by a number of different authors [2–11].

12.2   CALCULATION OF AWPCs FROM P_v AND SOLUBILITY PARAMETERS

For liquids with low water miscibility, AWPCs can be calculated as the ratio of the solute vapor pressure to the water solubility:

$$H_c(\text{atm} \cdot \text{m}^3 \, \text{mol}^{-1}) = \frac{p_v(\text{atm})}{C_w^s(\text{mol m}^{-3})} \qquad (12.2.1)$$

where both p_v and C_w^s have to be at the given temperature. Staudinger and Roberts [2] discuss this method in detail and specify two key assumptions: (1) the solubility of water in the organic liquid does not significantly affect the vapor pressure of the organic liquid, and (2) the activity coefficient does not vary appreciably with concentrations. The assumptions are typically met when the water solubility in the organic liquid is below 0.05 mol fraction or when the solubility of the organic liquid is low (<0.05 mol fraction) [8]. The compilations of AWPC data indicated above includes values derived using relation 12.2.1.

Replacing C_w by the infinite dilution activity coefficient in water, γ_w^∞, the following relation is obtained:

$$H_c(\text{atm} \cdot \text{m}^3 \, \text{mol}^{-1}) = 18 \times 10^{-6} \gamma_w^\infty (\text{m}^3 \, \text{mol}^{-1}) p_v(\text{atm}) \qquad (12.2.2)$$

where γ_w^∞ can be estimated from molecular structure input using the UNIFAC approach [12].

12.3   STRUCTURE–AWPC CORRELATION

Nirmalakhandan and Speece [13] developed a model for estimation of K_{aw} based on MCIs and a polarity term, Φ. This model is similar to the one used for estimating the water solubility (compare with eqs. 11.5.4 and 11.5.5). The model for estimating K_{aw}

includes an additional indicator variable I, accounting for electronegative elements (O, N, or halogen) attached directly to a hydrogen carrying a C atom. The model is

$$\log_{10} K_{aw} = 1.29 + 1.005\Phi - 0.468\,^1\chi^v + 1.258I$$
$$n = 180, \quad s = 0.262, \quad r = 0.99 \tag{12.3.1}$$

Φ is calculated with an additive atom and bond contribution scheme. The model applies for hydrocarbons, halogenated hydrocarbons, alcohols, and esters of alkanoic acids. Validation of eq. 12.3.1 with a test set of 20 compounds has been performed and discussed.

Aldehydes can become hydrated in water [i.e., establish an equilibrium between the hydrated (*gem*-diol form) and the unhydrated form]. Therefore, hydration should be considered in evaluating AWPCs. Betterton and Hoffmann [14] have investigated the correlation of AWPCs with Taft's parameter for substituted aldehydes.

12.4   GROUP CONTRIBUTION APPROACHES

Method of Hine and Mookerjee   Hine and Mookerjee [3] developed two models, a bond and a group model, for hydrocarbons, halogenated hydrocarbons, and compounds containing hydroxyl, ether, aldehydo, keto, carboxylic ester, amino, nitro, and mercapto groups. Their training set included data for 292 compounds which were either experimental or calculated from solubility and vapor pressure data. The training set used included only a few compounds with multiple group occurrence, such as pyrazines and dihydroxyl-, diamino-, and polyhalogenated compounds. *Distant polar interaction* terms apply for the latter compounds, to correct for the deviation from simple group additivity due to functional group interaction.

Method of Meylan and Howard   Meylan and Howard [9] expanded the bond contribution method of Hine and Mookerjee. Based on 345 compounds they derived bond contributions for 59 different bond types. Their method has been validated with an independent set of 74 structurally diverse compounds, obtaining a correlation coefficient of 0.96. Their method also needs correction factors for several structural–substructural features. This method has been implemented into a Henry's law constant program performing AWPC (25°C) estimations from SMILES input [15].

Method of Suzuki, Ohtagushi, and Koide   Suzuki et al. [16] developed a model to estimate K_{aw} at 25°C based on the MCI, $^1\chi$ (see Chapter 2), and group contributions:

$$\log_{10} K_{aw} = 0.40 - 0.34\,^1\chi + \sum n_i G_i \qquad n = 229, \quad s = 0.20, \quad r = 0.994 \tag{12.4.1}$$

The G_i values correspond to mono- or polyatomic groups characterized by their aliphatic or aromatic ring attachment. The training set used included data from the Hine and Mookerjee list upgraded with more recent data. Principal component analysis has been employed to propose $^1\chi$ as the most significant bulk structure

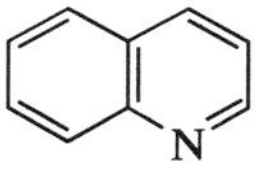

Quinoline

1. MCI: $^1\chi = 4.966$

2. Summation of contributions:

atom or group	$n_i G_i$	
C	9(−0.31)	− 2.79
H	7(0.30)	+ 2.10
N (aromatic)	1(−2.80)	− 2.80

$$\sum n_i G_i = -3.49$$

3. With eq. 12.4.1:

$$\log_{10} K_{aw} = 0.40 - 0.34(4.97) - 3.49$$
$$= -4.779$$
$$K_{aw} = 1.66 \times 10^{-5} \text{ at } 25°C$$

With eq. 12.1.3:

$$R = 82.06 \times 10^{-6} \text{ atm·m}^3(\text{mol·K})^{-1}$$
$$\rightarrow H_c = 1.66 \times 10^{-5}(82.06 \times 10^{-6})(298.2)$$
$$H_c = 4.06 \times 10^{-7} \text{ atm·m}^3\text{mol}^{-1} \text{ at } 25°C$$

Figure 12.4.1 Estimation of AWPC at 25°C for quinoline using the method of Suzuki et al. [16].

descriptor. Functional group contributions are the most significant descriptors for molecular cohesiveness and polarity. Equation 12.4.1 can be regarded as a specification of the following general approach:

$$\log_{10} K_{aw} = \text{constant} + \text{bulk structure} + \text{cohesiveness} + \text{polarity-related factors}$$

$$(12.4.2)$$

Model 12.4.1 has been implemented in the Toolkit. An application is shown for quinoline in Figure 12.4.1. An experimental $\log_{10} K_{aw}$ value of − 4.170 is known for quinoline [9].

12.5   TEMPERATURE DEPENDENCE OF AWPC

The temperature dependence of AWPC often follows the equation

$$\log_b \text{AWPC} = A - \frac{B}{T} \qquad (12.5.1)$$

where $\log_b$ is either $\log_{10}$ or ln, AWPC is either K_{aw}, H_c, or H_y, T is the absolute temperature in K, and A and B temperature coefficients. This equation can be derived assuming that the AWPC obeys the van't Hoff equation [2]. A compilation of temperature functions along with the applicable temperature ranges AWPC is given for hydrocarbons and halogenated hydrocarbons in Tables D.1 through D.8 in

Appendix D and in the compilation provided by Staudinger and Roberts [2]. The property estimation Toolkit by Reinhard and Drefahl includes selected AWPC temperature functions. An example is shown for 1,1,2-trichloroethane at 15°C in Figure 12.5.1 using the temperature coefficients measured by Ashworth et al. [17] and in Figure 12.5.2 using the coefficients of Leighton and Calo [18].

A compound with given temperature coefficients A and B is considered at two different temperatures: (1) a reference temperature, T_{ref}; and (2) an arbitrary temperature, T_x. Using eq. 12.5.1, K_{aw} at these temperatures is given by

$$\ln K_{aw}(T_{ref}) = A + BT_{ref}^{-1} \tag{12.5.2a}$$

$$\ln K_{aw}(T_x) = A + BT_x^{-1} \tag{12.5.2b}$$

Subtraction of (12.5.2a) from (12.5.2b) leads to

$$\ln K_{aw}(T_x) = \ln K_{aw}(T_{ref}) - BT_{ref}^{-1} + BT_x^{-1} \tag{12.5.3}$$

Note that for sets of compounds with equal B, application of eq. 12.5.3 requires solely the knowledge of K_{aw} at one reference temperature T_{ref}, that is, eq. 12.5.2 allows the K_{aw} extrapolation to an arbitrary temperature T_x irrespective of the particular molecular structure of a given compounds as long as the compound belongs to the given set.

Next we consider the problem where K_{aw} is needed at a certain temperature T_x and AWPC–temperature functions such as 12.5.2 are not available for the compound of interest, but K_{aw} is known for at least one reference temperature T_{ref}.

Cl H
| |
Cl—C—C—H
| |
H   Cl

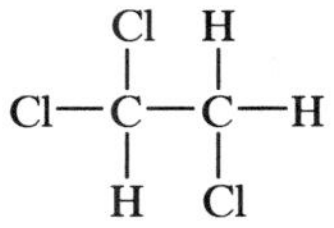

1,1,2-Trichloroethane

1. Temperature coefficient: $A = 9.320$, $B = 4843$ [17]
Range: 10–30°C
Units: H_c in atm·m^{3}mol^{-1}, T in K, $\log_b = \ln$

2. With eq. 12.5.1: $\ln H_c = 9.320 - \dfrac{4843}{288.2}$

$$= -7.484$$
$$H_c = 0.000562 \text{ atm·m}^3\text{mol}^{-1} \text{ at } 15°C$$

With eq. 12.1.2: $R = 82.06 \times 10^{-6} \text{ atm·m}^3(\text{mol·K})^{-1}$

$$\rightarrow K_{aw} = \frac{0.000562}{82.06 \times 10^{-6}(288.2)}$$

$$K_{aw} = 0.0238 \text{ at } 15°C$$

Figure 12.5.1 Estimation of AWPC at 15°C for 1,1,2-trichloroethane [data from Ref. 17].

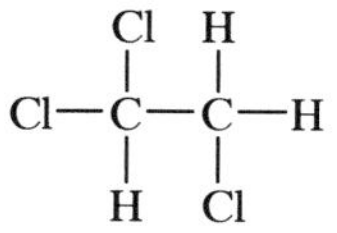

$$\begin{array}{cc}
\text{Cl} & \text{H} \\
| & | \\
\text{Cl}-\text{C}-\text{C}-\text{H} \\
| & | \\
\text{H} & \text{Cl}
\end{array}$$

1,1,2-Trichloroethane

1. Temperature coefficient: A = 16.20, B = 3690 [18]
Range: 0–30°C
Units: H_y [–], T in K, $\log_b = \ln$

2. With eq. 12.5.1: $\ln H_y = 16.20 - \dfrac{3690}{288.2}$

$= 3.396$
$H_y = 29.86$ at 15°C

With eq. 12.1.3: $R = 82.06$ atm·m^3(mol K)$^{-1}$, $p = 1$ atm
Assumptions: (1) Solution density equals water density
(2) Density temperature independent
then: $V_s = 18 \times 10^{-6}$ m^3 mol^{-1}

$\rightarrow K_{aw} = 29.86(1)\left[\dfrac{18 \times 10^{-3}}{82.06 \times 10^{-6}(288.2)}\right]$

$K_{aw} = 0.0227$ at 15°C

With eq. 12.1.2 $R = 82.06$ atm·m^3(mol K)$^{-1}$
$\rightarrow H_c = 0.0277(82.06 \times 10^{-6})(288.2)$
$H_c = 0.000655$ atm·m^3 mol^{-1} at 15°C

Figure 12.5.2 Estimation of AWPC at 15°C for 1,1,2-trichloroethane [data from Ref. 18].

This problem can be solved knowing that for certain sets of compounds such as the trihalomethanes (THMs), B is approximately independent of the compound's particular structure; that is, plots of $\ln K_{aw}$ versus T^{-1} show a set of parallel lines. Given that B is nearly constant within in the temperature range, and substituting B in eq. 12.5.3 by C_1 and BT_{ref}^{-1} by C_0 leads to

$$\ln K_{aw}(T_x) = \ln K_{aw}(T_{ref}) - C_0 + C_1 T_x^{-1} \qquad (12.5.4)$$

Based on the assumption above, C_0 and C_1 are compound-independent constants. Equation 12.5.4 has been tested for the four reference temperatures 15, 20, 25, and 30°C using temperature-dependent K_{aw} data derived with the coefficients A and B in Appendix D. The coefficients C_0 and C_1 have been derived by linear regression [$\ln K_{aw}(T_x) - \ln K_{aw}(T_{ref})$] versus T_x^{-1}, presented in Table 12.5.1 along with the statistical parameters. Equation 12.5.4 is useful in estimating K_{aw} at a temperature of interest if K_{aw} is known at any of the reference temperatures. Since B is not truly constant, estimation of $K_{aw}(T_x)$ is associated with an error that increases with increasing difference between T_{ref} and T_x. Thus the estimation should be performed with the K_{aw} at the closest available T_{ref} if experimental K_{aw} values at more than one temperature are available.

TABLE 12.5.1   Temperature coefficients and Statistical Parameters for Eq. 12.5.4 at Four Different Reference Temperatures

T_{ref} (K)	C_0	C_1	s^a	r^b
288.15	14.648	4220.9	0.178	0.9996
293.15	14.105	4135.0	0.168	1.0000
298.15	13.874	4136.6	0.163	1.0000
303.15	12.592	3817.4	0.145	0.9996

^{a} Standard deviation.
^{b} Correlation coefficient.

REFERENCES

1. Munz, C., and P. V. Roberts, Gas- and Liquid-Phase Mass Transfer Resistance of Organic Compounds During Mechanical Surface Aeration. *Water Res.*, 1989: **23**, 589–601.

2. Staudinger, J., and P. V. Roberts, A Critical Review of Henry's Law Constants for Environmental Applications. *Crit. Rev. Environ. Sci. and Techno.*, 1996: **26**(3), 205–297.

3. Hine, J., and P. K. Mookerjee, The Intrinsic Hydrophilic Character of Organic Compounds: Correlations in Terms of Structural Contributions. *J. Org. Chem.*, 1975: **40**, 292–298.

4. Mackay, D., and W. Y. Shiu, A Critical Review of Henry's Law Constants for Chemicals of Environmental Interest. *J. Phys. Chem. Ref. Data*, 1981: **10**, 1175–1199.

5. Shiu, W. Y., and D. A. Mackay, A Critical Review of Aqueous Solubilities, Vapor Pressures, Henry's Law Constants, and Octanol–Water Partition Coefficients of the Polychlorinated Biphenyls. *J. Phys. Chem. Ref. Data*, 1986: **15**(2), 911–926.

6. Shiu, W. Y., et al., Physical–Chemical Properties of Chlorinated Dibenzo-*p*-dioxins. *Environ. Sci. Technol.*, 1988: **22**, 651–658.

7. Suntio, L. R., W. Y. Shiu, and D. Mackay, A Review of the Nature and Properties of Chemicals Present in Pulp Mill Effluents. *Chemosphere*, 1988: **17**, 1249–1290.

8. Suntio, L. R., et al., Critical Review of Henry's Law Constants for Pesticides. *Rev. Environ. Contam. Toxicol.*, 1988: **103**, 1–59.

9. Meylan, W. M., and P. H. Howard, Bond Contribution Method for Estimating Henry's Law Constants. *Environ. Toxicol. Chem.*, 1991: **10**, 1283–1293.

10. Yaws, C., H. C. Yang, and X. Pan, Henry's Law Constants for 362 Organic Compounds in Water. *Chem. Eng.*, 1991: Nov., 179–185.

11. Lucius, J. E., et al., Properties and Hazards of 108 Selected Substances, 1992. U. S. Geological Survey Open File Report. Washington, DC: U.S. Geological Survey.

12. Arbuckle, W. B., Estimating Activity Coefficients for Use in Calculating Environmental Parameters. *Environ. Sci. Technol.*, 1983: **17**, 537–542.

13. Nirmalakhandan, N. N., and R. E. Speece, QSAR Model for Predicting Henry's Constant. *Environ. Sci. Technol.*, 1988: **22**, 1349–1357.

14. Betterton, E. A., and M. R. Hoffmann, Henry's Law Constants of Some Environmentally Important Aldehydes. *Environ. Sci. Technol.*, 1988: **22**, 1415–1418.

15. Meylan, W. M., and P. H. Howard, *Henry's Law Constant Program*, 1992. Boca Raton, FL: Lewis Publishers.

16. Suzuki, T., K. Ohtaguchi, and K. Koide, Application of Principal Components Analysis to Calculate Henry's Constant from Molecular Structure. *Comput. Chem.*, 1992: **16**, 41–52.

17. Ashworth, R. A., et al., Air–Water Partitioning Coefficients of Organics in Dilute Aqueous Solutions. *J. Hazard. Mat.*, 1988: **18**, 25–36.

18. Leighton, D. T. J., and J. M. Calo, Distribution Coefficients of Chlorinated Hydrocarbons in Dilute Air–Water Systems for Groundwater Contamination Applications. *J. Chem. Eng. Data*, 1981: **26**, 382–385.

19. Nicholson, B. C., B. P. Maguire, and D. B. Bursill, Henry's Law Constants for the Trihalomethanes: Effect of Water Composition and Temperature. *Environ. Sci. Technol.*, 1984: **18**, 518–521.

CHAPTER 13

1-OCTANOL–WATER PARTITION COEFFICIENT

13.1 DEFINITIONS AND APPLICATIONS

The 1-octanol–water partition coefficient, K_{ow}, is defined as

$$K_{ow} = \frac{C_o}{C_w} \qquad (13.1.1)$$

where C_o and C_w refer to the molar, or mass, concentrations in the water-saturated octanol and in the octanol-saturated water phase, respectively [1, 2]. K_{ow} is often abbreviated as P or P_{ow} and $\log_{10} K_{ow}$ can be used as a relative measure of a compound's hydrophobicity. The hydrophobicity scale ranges from -2.6 for hydrophilic compounds such as 4-aminophenyl β-D-glucopyranoside [3] to $+8.5$ for hydrophobic compounds such as decabromobiphenyl [4]. The $\log_{10} K_{ow}$ value has been termed the Hansch parameter [5].

USES FOR 1-OCTANOL/WATER PARTITIONING DATA

- To estimate soil–water partition coefficients
- To estimate dissolved organic matter–water partition coefficients
- To estimate lipid solubility [6]
- To estimate bioconcentration factors
- To estimate aqueous toxicity parameters
- To estimate biodegradation parameters
- To assess the formation of micelles [7]

TABLE 13.1.1 Applicable K_{ow} Range of Different Experimental Method

Method Type	Method	K_{ow} Range
Direct	Shake flask	-2.5 to 4.5
	Slow stirring	<8
	Generator column	<8.5
Indirect	Reversed-phase HPLC	0 to 6
	Reversed-phase TLC	0 to 12

Source: Adapted from Ref. [8].

Experimental Method The choice of method for experimental K_{ow} determination depends on the K_{ow} value of the target compound. Table 13.1.1 presents common experimental methods with applicable K_{ow} ranges [8]. Values may differ depending on the method used (for a discussion see refs. [4, 6, 8]).

Dependence on Temperature For most compounds, values of K_{ow} have been determined at temperatures between 20 and 25°C. The effect of temperature on K_{ow} is small, usually lower than $\pm 0.02 \log_{10} K_{ow}$ unit per degree. For example, log K_{ow} of 1,2,3,4-chlorobenzene has been determined with the slow stirring method as 4.635 ± 0.004 at 25°C [9] and as 4.564 ± 0.074 at 29°C [8].

Dependence on Solution pH Solution conditions can affect K_{ow} values. For an organic acid HA that ionizes to A^- at the solution pH of interest, the overall K'_{ow} is given by

$$K'_{ow} = \{[\text{HA}]_o + [\text{A}^-]_o\}/\{[\text{HA}]_{aq} + [\text{A}^-]_{aq}\} \qquad (13.1.2)$$

where the squared brackets indicate compound concentration and the subscripts o and aq refer to the octanol and aqueous phase, respectively. The K_{ow} of unionized (molecular) HA is expected to be significantly greater than that of A^-, which has a greater affinity for the aqueous phase. For acids, K'_{ow} is a function of the fraction of HA present, ϕ_{HA}. ϕ_{HA} is defined as

$$\phi_{\text{HA}} = [\text{HA}]/\{[\text{HA}] + [\text{A}^-]\} \qquad (13.1.3)$$

ϕ_{HA} is a function of the proton concentration, $[\text{H}^+]$, and the acid dissociation constant, K_{HA}.

$$\phi_{\text{HA}} = [\text{H}^+]/\{[\text{H}^+] + K_{\text{HA}}\} \qquad (13.1.4)$$

The overall K'_{ow} is then obtained from:

$$K'_{ow} = K_{ow,\text{HA}}\phi_{\text{HA}} + K_{ow,\text{A}}(1 - \phi_{\text{HA}})$$

where $K_{ow,\text{HA}}$ and $K_{ow,\text{A}}$ are the K_{ow} values for the molecular and the ionized forms of HA, respectively.

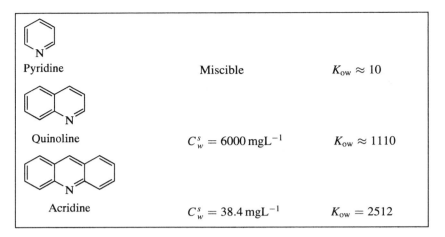

Figure 13.2.1 K_{ow} versus C_w^s for N-containing aromatic heterocyclic compounds [10].

13.2   PROPERTY–K_{ow} CORRELATIONS

Solubility–K_{ow} Correlations   The decrease of aqueous solubility results into increased hydrophobicity. This qualitative, inverse relationship is demonstrated for N-containing aromatic heterocyclic compounds in Figure 13.2.1.

Quantitative relationships between K_{ow} and C_w^s have been reviewed by Isnard and Lambert [11]. These authors developed a model based on 300 structurally diverse compounds. The model equation for liquids ($T_m < 25°C$) is

$$\log_{10}K_{ow} = 3.06 - 0.68 \log_{10}[C_w^s(\text{mol m}^{-3})] \qquad s = 0.466, \quad r = 0.965 \quad (13.2.1)$$

and for solids ($T_m > 25°C$) the equation is

$$\log_{10}K_{ow} = 3.15 - 0.72\{\log_{10}[C_w(\text{mol m}^{-3})] - 0.0038[T_m(°C) - 25°C]\}$$
$$s = 0.440, \quad r = 0.965 \qquad (13.2.2)$$

Bowman and Sans [12] measured K_{ow} and water solubilities at 20°C for liquid and solid carbamates and organophosphorous insecticides and related compounds. Based on these data, they derived the following equation:

$$\log_{10}K_{ow} = 0.323 - 0.833\{\log_{10}[C_w^s(\text{mol L}^{-3})]_{\text{corr}}\} \qquad n = 58, \quad r = -0.975$$
$$(13.2.3)$$

where $\log_{10}(C_w^s)_{\text{corr}}$ is either the liquid solubility or the melting point–corrected solubility of the solid:

$$\log_{10}(C_w^s)_{\text{corr}} = \log_{10}(C_w^s)_{\text{solid}} + \frac{\Delta H_{\text{fus}}}{2.303R} \frac{T_m - T}{TT_m} \qquad (13.2.3a)$$

where T_m and T are in K, ΔH_{fus} is the heat of fusion in cal mol^{-1}, and R is the universal gas constant (1.98717 cal K^{-1} mol^{-1}). Since no accurate ΔH_{fus} values were

TABLE 13.2.1   Coefficients and Statistical Parameters in Model 13.2.1 [13]

Liquid class	a_0	a_1	n	r
n-Alkanes	-0.468 ± 0.081	0.972 ± 0.016	4	0.999
1-Alkenes and alkynes	-0.250 ± 0.105	0.908 ± 0.063	6	0.993
Subst. benzenes[a]	-0.768 ± 0.100	1.056 ± 0.026	18	0.995
Halogenated hydrocarbons[b]	-0.323 ± 0.133	0.907 ± 0.033	13	0.993
1-Alkanols[c]	-0.348 ± 0.112	1.030 ± 0.011	6	0.997
Aldehydes, ketones[d]	-0.465 ± 0.155	1.079 ± 0.065	8	0.989
Alkanoates[e]	-0.285 ± 0.167	0.932 ± 0.005	7	0.991
All compounds	-0.311 ± 0.066	0.944 ± 0.018	62	0.990

[a] Substituted benzenes alkyl, fluoromethyl, chloro, iodo, hydroxy, nitro.
[b] 1-Chloro- and 1-bromoalkanes, 1-iodoheptane, mono- and polyhalogenated alkenes.
[c] + 2-Ethyl-1,3-hexanediol.
[d] 2-Alkanones, 3-pentanone, acetal, 2-furaldehyde.
[e] + 2-Bromoethyl ethanoate.
Source: Ref. 13. Reprinted with permission. Copyright (1982) American Chemical Society.

available for most biocides, $\Delta H_{fus}/T_m = 13.5 \pm 3$ eu was found to be a reasonably accurate estimate for the entropy of fusion of most-low-melting solids [12].

Tewari et al. [12] measured K_{ow} and C_w^s for 62 liquid compounds and derived the following relationship for each compound class and for all compounds together:

$$\log_{10} K_{ow} = a_0 + a_1 [C_w^s (\text{gm}^{-3}) V_M (\text{m}^3 \, \text{g}^{-1})]^{-1} \qquad (13.2.4)$$

where all properties are at 25°C. The derived regression coefficients a_0 and a_1 and the statistical parameters are listed in Table 13.2.1. K_{ow} can be estimated with reasonable accuracy for liquids of these classes. However, in addition to C_w^s, V_M has to be known.

Activity Coefficient–K_{ow} Relationships K_{ow} can be estimated knowing the activity coefficients for the aqueous and the octanol phase:

$$K_{ow} = \frac{0.115 \gamma_w^\infty}{\gamma_o^\infty} \qquad (13.2.5)$$

where γ_w^∞ and γ_o^∞ are the infinite dilution activity coefficients for the solute in water and in octanol, respectively [14]. For certain compounds, the activity coefficients can be estimated using the UNIFAC model [15].

Collander-Type Relationships   Collander has studied partition coefficients in different alcohol–water systems [16]. He found that these partition coefficients are mutually correlated. For certain compounds containing one hydrophilic group, such as alkanols, alkanoic acids, alkanoates, dialkyl ethers, and alkylamines and selected compounds containing two, three, or four such groups, he reports the following equation:

$$\log_{10} K_{ow} = -0.42 + 1.24 \log_{10} K_{\text{butanol/W}} \qquad (13.2.6)$$

where $K_{butanol/w}$ is the 1-butanol/water partition coefficient. Collander discusses the distinct difference of molecules with respect to their number of hydrophilic groups. Equation 13.2.6 slightly underestimates K_{ow} for the monohydrophilic, whereas K_{ow} for polyhydrophilic compounds is overestimated.

Muller's Relationship   Muller [17] derived the following collander-type relationship for the monohydrophilic class of alkanols (C$_2$–C$_6$):

$$\log_{10}K_{ow} = -0.14 + 1.210\log_{10}K_{benzylalcohol/w} \qquad (13.2.7)$$

where $K_{benzylalcohol/w}$ is the benzylalcohol–water partition coefficient. Analogous relationships between K_{ow} and organic solvent–water partition coefficients have been reviewed by Lyman [2].

LSER Approach   The LSER approach has been described for aqueous solubility in Section 11.4. He et al. [18] have derived the following relationship for phenyl-sulfonyl alkanoates:

$$\log_{10}K_{ow} = (1.833 \pm 0.567) + \frac{(3.111 \pm 0.109)V_i}{100} - (0.397 \pm 0.156)\pi^*$$
$$- (3.162 \pm 0.277)\beta \qquad n = 28, \quad s = 0.102, \quad r = 0.986 \quad (13.2.8)$$

where K_{ow} is at 25°C, and V_i, π^*, and β are the solvatochromic parameters (see chapter 11.4)

Chromatographic Parameter–K_{ow} Relationships   Correlations between K_{ow} and various chromatographic parameters (CGP), such as HPLC retention time and thin-layer chromatography (TLC) capacity factors, allow the experimental estimation of K_{ow} [19]. Usually, the CGP–K_{ow} correlation is evaluated for a calibration set of compounds with accurately known K_{ow} values. The K_{ow} of a new compound can then be estimated by determining its CGP under the same experimental conditions as those used for the calibration set.

Veith et al. [20] studied the correlation between K_{ow} and (C$_{18}$) reversed-phase HPLC retention time for a wide variety compounds, such as substituted benzenes, PAHs, and PCBs. Their calibration set consisted of benzene, bromobenzene, biphenyl, bibenzyl, DDE, and 2,2',4,5,5'-pentachlorobiphenyl with measured log K_{ow} values of 2.13, 2.99, 3.76, 4.81, 5.69, and 6.11, respectively. Similarly, Chin et al. [21] reported a relationship using phenol, nitrobenzene, toluene, chlorobenzene, naphthalene, *o*-xylene, *o*-dichlorobenzene, 1,2,4-trichlorobenzene, biphenyl, and anthracene as a calibration set where log K_{ow} ranges from 1.46 to 4.54. Burkhard et al. [22] performed an analogous study using chlorobenzenes and PCBs as the calibration set with log K_{ow} ranging from 2.62 to 8.23. McDuffie [23] discussed a relationship to estimate K_{ow} for various halogenated hydrocarbons and pesticides. Rapaport and Eisenreich [24] reported a HPLC retention time–K_{ow} relationship exclusively for PCBs. They discussed the relationship with respect to different substitution patterns in isomeric PCBs. Average log K_{ow} vlaues for isomeric classes range from 4.5 for $N_{Cl} = 1$ to 8.1 for $N_{Cl} = 7$. Karcher [25] reports similar relationships for chlorinated dibenzo-*p*-dioxins and dibenzofurans.

Biagi et al. [28] studied the relationship between K_{ow} and reversed-phase TLC retention factor for 28 phenols substituted with alkyl, halogen, methoxy, and nitro groups. Budvári-Bárány et al. [29] compare the HPLC and TLC retention method to estimate K_{ow} for a class of heterocyclic compounds (imidazoquinoline derivatives). Takács-Novák et al. [30] has demonstrated the similarity between the pH-dependent K_{ow} and the pH-related retention (C_{18} / methanol–water) pattern of eight amphoteric compounds in the pH range 4 to 9.

13.3 STRUCTURE–K_{ow} RELATIONSHIPS

Graph-theoretical invariants such as MCIs and ICIs account for size and shape characteristics in molecules of organic compounds. Functional characteristics such as the hydrogen-bonding capability have to be incorporated into structure–property models by either indicator variables or by structuring the model as a set of separate equations where each equation applies for a structurally specified compound class. The latter approach has been investigated by Niemi et al. [31] in modeling K_{ow} as a function of various graph-theoretical invariants for classes of compounds, each containing compounds with equal numbers of hydrogen bonds. The authors conclude that their models are not necessarily an alternative to currently available methods such as K_{ow} GCMs. They stress, however, the advantage of their type of models with respect to the error-free calculability of graph-theoretical invariants for any arbitrary structure.

In the following, compound-class-specific correlations between K_{ow} and selected molecular descriptors such as chlorine number, molecular connectivity indices, van der Waals volume and area, molecular volume, and polarizability are reviewed. Further, the model of Bodor, Babanyi, and Wong will be introduced, which allows estimation from molecular structure input for a broad range of compounds.

Chlorine Number–K_{ow} Relationships For chlorinated aromatic compounds, linear correlations between K_{ow} and N_{Cl} have been reported [32]:

$$\log_{10} K_{ow} = a_0 + a_1 N_{Cl} \tag{13.3.1}$$

Coefficients a_0 and a_1 and the statistical parameters derived for three classes for compounds are presented in Table 13.3.1. Model 13.3.1 allows order of magnitude estimations for K_{ow} but does not account for particular substitution patterns.

TABLE 13.3.1 Coefficients and Statistical Parameters in model 13.3.1

Compound Class	a_0	a_1	n	s	r^2
Chlorobenzenes	2.30	0.58	13	0.07	0.996
Chloroanilines	1.10	0.85	14	0.11	0.992
PCBs	4.36	0.45	20	0.29	0.935
PCDDs[a]	4.35	0.65	—	—	—

[a] For $N_{Cl} \leq 4$ [33].
Source: Refs. 31 and 32.

Molecular Connectivity–K_{ow} Relationships   Kier and Hall [34] have analyzed the correlation between K_{ow} and various MCIs for hydrocarbons and monofunctional alcohols, ethers, ketones, acids, esters, and amines. Analogous relationships have been studied by Finizio et al. [35] for substituted s-triazines and by Govers et al. [36] for thioureas.

Doucette and Andren [4] have compared six methods to estimate K_{ow} for highly hydrophobic aromatic compounds such as halogenated benzenes, biphenyls, dibenzofurans, and dibenzo-p-dioxins with log K_{ow} values ranging from 2.13 to 8.58. The comparison includes the GCM of Hansch and Leo, the GCM of Nys and Rekker, and correlations based on the following molecular descriptors: HPLC retention times, M, TSA, and MCIs. The method using MCIs had the smallest average percent error. The method is

$$\log_{10} K_{ow} = -0.085 + 1.27\,^1\chi^v - 0.050(^1\chi^v)^2 \qquad n = 64, \quad r^2 = 0.967 \quad (13.3.2)$$

Basak et al. [37] derived the following model for non-hydrogen-bonding compounds such as alkanes, alkylbenzenes, PAHs, chlorinated alkanes, chlorobenzenes, and PCBs:

$$\log_{10} K_{ow} = -3.127 - 1.644\,IC_0 + 2.120\,^5\chi_C - 2.9140\,^6\chi_{CH} + 4.208\,^0\chi^v$$
$$+ 1.060\,^4\chi^v - 1.020\,^4\chi^v_{PC} \qquad n = 137, \quad s = 0.26, \quad r^2 = 0.97$$
$$(13.3.3)$$

where IC_0 is the zero-order information content (e.q., 2.3.10).

Characteristic Root Index–K_{ow} Relationships   Saçan and Inel [38] derived a relationship between K_{ow} and the characteristic root index (CRI) for PCBs (Cl_0–Cl_8):

$$\log_{10} K_{ow} = 1.330 + 1.068\,CRI \qquad n = 34, \quad S_{y,x} = 0.116, \quad r = 0.997,$$
$$AD = 0.08 \qquad (13.3.4)$$

Only K_{ow} data of PCBs with more than one experimental value have been included in the regression. The $\log_{10} K_{ow}$ data range from 3.89 for biphenyl to 8.28 for 2,2′, 3,3′,5,5′,6,6′-octachlorobiphenyl.

Extended Adjacency Matrix–K_{ow} Relationships   Yang et al. [39] derived two descriptors, EA_Σ and EA_{max}, from the extended adjacency (EA) matrix and demonstrated their correlation with $\log_{10} K_{ow}$ for barbiturate acid derivatives with structures I and II:

$$\log_{10} K_{ow} = -2.8302 + 0.1900\,EA_\Sigma - 0.7395\,EA_{max} \qquad n = 25(n_I = 14, n_{II} = 11),$$
$$s = 0.0861, \quad R = 0.9910, \quad F = 602.9 \qquad (13.3.5)$$

I   II

Van der Waals Parameter–K_{ow} Relationships   Moriguchi et al. [40] have studied the correlation of K_{ow} with either the van der Waals volume, V_{vaw}, or the van der Waals surface area, A_w, for apolar organic compounds such as inert gases, alkylbenzenes, PAHs, and halogenated alkanes and benzenes. The relationships are

$$\log_{10}K_{ow} = 0.23(\pm 0.15) + 2.51(\pm 0.13)V_{vaw}(\text{Å}^3) \qquad n = 60, \quad s = 0.228,$$
$$r = 0.980 \tag{13.3.6a}$$

$$\log_{10}K_{ow} = 0.04(\pm 0.21) + 2.17(\pm 0.15)A_w(\text{Å}^2) \qquad n = 60, \quad s = 0.295,$$
$$r = 0.966 \tag{13.3.6b}$$

The van der Waals parameters are calculated from atom and bond contributions provided by Moriguchi et al. [40] who give a sample calculation for bromopropane.

Molecular Volume–K_{ow} Relationships   Relationships between K_{ow} and different volume parameters have been reported. Leo et al. [41] compare correlations with Bondi and with CPK volume for two classes of apolar molecules: (1) alkanes and alkylsilanes, and (2) perhalogenated alkanes and aromatic and haloaromatic compounds. Further, these authors discuss analogous correlations for alkanols and alkylphenols.

Polarizability–K_{ow} Relationships   Molar polarizabilities can be derived from molecular orbital (MO) calculations by the complete neglect of differential overlap (CNDO) method [42]. The following correlation has been found for polar compounds that contain either hydrogen-bond-accepting or hydrogen-bond-donating groups (alkanols, alkanones, dialkyl ethers, alkanenitriles):

$$\log_{10}K_{ow} = (-1.60 \pm 0.17) + (0.27 \pm 0.01)\alpha_{CNDO/2}(\text{Å}^3) \qquad n = 14, \quad s = 0.267,$$
$$r = 0.961, \quad F = 144.9 \tag{13.3.7}$$

where $\alpha_{CNDO/2}$ is the volume polarizability obtained by summation over CNDO/2-derived atom polarizabilities in the molecule [42]. The authors also discuss correlations between K_{ow} and the dipole moment, μ, and the energy of the highest occupied orbital, $E(\text{HOMO})$.

Model of Bodor, Gabanyi, and Wong   The model Bodor et al. [43] is based on a nonlinear correlation of $\log K_{ow}$ with molecular descriptors such as molecular surface, volume, weight, and charge densities on N and O atoms in the molecule. The equation is

$$\log_{10}K_{ow} = 27.273 - 1.167 \times 10^{-4}S^2 - 6.106 \times 10^{-2}S + 14.87O^2 - 43.67O$$
$$+ 0.9986\,I_{alkane} + 9.57 \times 10^{-3}M - 0.1300\,D - 4.929\,Q_{ON} - 12.17Q_N^4$$
$$+ 26.81\,Q_N^2 - 7.416\,Q_N - 4.551\,Q_O^4 + 17.92\,Q_O^2 - 4.03\,Q_O \quad n = 118,$$
$$s = 0.296, \quad r = 0.939, \quad F^2 = 115.1 \tag{13.3.8}$$

where S is the molecular surface, O the ovality of the molecule, I_{alkane} the indicator for alkanes (its value is 1 for alkanes, otherwise 0), M the molar mass, D the calculated dipole moment, Q_{ON} the sum of absolute values of atomic charges on N and O atoms, Q_{N} is the square root of sum of squared charges on N atoms, and Q_{O} is the square root of sum of squared charges on O atoms. The training set includes alkanes, alkylbenzenes, halogenated alkanes, alkanols, ether, alkanones, alkanoic acids, esters, amines, nitriles, and selected multifunctional and hetereocyclic compounds in the log K_{ow} range between -1.31 and 4.06.

Artificial Neural Network Model of Bodor, Huang, and Harget   Bodor et al. [44] have studied the utility of an ANN to predict K_{ow} from quantum-chemically derived descriptors. Model training was performed with 302 compounds. The model was tested with 21 compounds not included in the training set. The authors compared the ANN approach with the regression analysis approach and concluded that ANN results compared favorably with those given by the regression model for both the training set and the test compounds.

13.4   GROUP CONTRIBUTION APPROACHES FOR K_{ow}

The Methylene Group Method of Korenman, Gurevich, and Kulagina
Korenman et al. [45] studied the solvent–water partition coefficients K_{sw} of ethylamine, *n*-propylamine, and *n*-butylamine at 20°C. Based on the data for these amines with a short, linear alkyl chain, the contributions to the logarithm of the solvent–water partition coefficients were found to be constant among particular solvent classes. The contribution values are shown in Table 13.4.1. Accordingly, $\Delta \log K_{\text{ow}}$ for CH_2 is 0.50.

Method of Broto, Moreau, and Vandycke   The method of Broto et al. [46] is an atom contribution method including one extra contribution for conjugated double bonds. The complete set of atom constants is given in Appendix F to illustrate the relative hydrophobicity of the different types of atom contributions. Atom types are differentiated by their environment depending on whether they are C atoms or heteroatoms. The C-atom environment is limited to the adjacent bonds and to the attached H atoms. For heteroatoms, the environment additionally includes nonhydrogen neighbors. The latter are divided into two classes: (1) C atoms, for which the bond environment is considered; and (2) heteroatoms, Z, irrespective of

TABLE 13.4.1 $-CH_2-$ Group Contribution to the Solvent–Water Partition Coefficient for Various Solvent Classes

Solvent Class	$\Delta \log k_{\text{sw}}(-CH_2-)$
n-Alkanols (C_4–C_{10})	0.50 ± 0.02
n-Alkanes (C_5–C_{10}), cyclohexane, tetrachloromethane	0.41 ± 0.02
Benzene, alkylbenzenes, mono-, di-, and tri-halobenzenes	0.51 ± 0.02
Trichloromethane, 1,2-dichloroethane	0.65 ± 0.03

Source: Ref. [45].

–C– C in methylene group	–C≡ C as in alkynyl or nitrilo group	N≡C– N in nitrilo group	F–C≡ F as in 1-fluoroalkyne	F–Z F as in fluoroamino group

Figure 13.4.1 Four different atom groups as defined in the model of Broto et al. [46].

their particular environment. For example, the C atom in group **F–C≡** of Figure 13.4.1 is specified by both the single bond to F, and the triple bond that connects this group with the remainder of the molecule. In contrast, the Z atom in group **F–Z** is characterized by the single bond to F irrespective of how Z is connected to the remainder molecule. **F–Z** could be a fluoroamino group with two adjacent single bonds or a P-containing group where different combination of double and single bonds are possible.

This model encompasses 222 atom contributions derived from a training set with 1868 experimental log K_{ow} values. Compounds capable of intramolecular hydrogen bonding were excluded from the training set. Experimental measurements done in acidic or buffered solutions were not used. K_{ow} estimations are estimated to be precise to within 0.4 log unit (equivalent to the precision of many experimental determinations). The model is implemented in the Toolkit and in SmilogP [42]. The application of this method to two compounds, 2-bromoethyl ethanoate and 2,4-dinitro-6-*sec*-butylphenol, is illustrated in Figures 13.4.2 and 13.4.3. Experimental log K_{ow} values for these compounds are 1.11 for 2-bromoethyl ethanoate [13] and 3.143 ± 0.010 (also: 3.69; 4.1 ± 0.2) for 2,4-dinitro-6-*sec*-butylphenol [9].

Intramolecular hydrogen bonding increases compound hydrophobicity. Broto et al. evaluated the effect of intramolecular hydrogen bonding on compound lipophilicity by comparing the observed log K_{ow} and the calculated K_{ow} of 500 compounds capable of forming intramolecular hydrogen bonds. On average the observed log K_{ow}

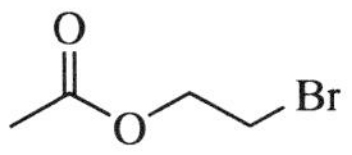

2-Bromoethyl ethanoate

Atomic group	Location	Contribution terms	
C–(No. 3)	ethanoate group	10.631	0.631
⟩**C**= (No. 13)	ethanoate group	1(−0.548)	−0.548
O=C⟨ (No. 14)	ethanoate group	1(−0.681)	−0.681
–**O**–C(=)– (No. 62)	ethanoate group	1(0.201)	0.201
–**C**– (No. 2)	ethyl group	2(0.456)	0.912
Br–C– (No. 44)	bromo group	1(0.620)	0.620
		$\log_{10} K_{ow}$ =	1.135
		K_{ow} =	13.65

Figure 13.4.2 Estimation of K_{ow} for 2-bromoethyl ethanoate using the method of Broto et al. [46].

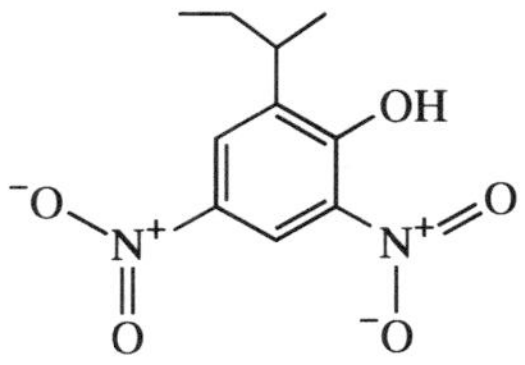

2,4-Dinitro-6-*sec*-butylphenol

Atomic group	Location	Contribution terms	
c–N(=Z)–Z (No. 176)	nitro group	2(0.284)	0.568
O=Z (No. 33)	nitro group	4(0.339)	1.356
O–c (No. 20)	hydroxy group	1(−0.173)	−0.173
C– (No. 3)	*sec*-butyl group	2(0.631)	1.262
–C– (No. 2)	*sec*-butyl group	1(0.456)	0.456
–C(−)– (No. 4)	*sec*-butyl group	1(0.029)	0.029
··cH·· (No. 6)	benzene ring	2(0.311)	0.622
··C(−)·· (No. 5)	benzene ring	4(0.095)	0.380

$$\log_{10} K_{ow} = 4.500$$
$$K_{ow} = 3.16 \times 10^4$$

Figure 13.4.3 Estimation of K_{ow} for 2,4-dinitro-6-*sec*-butylphenol using the method of Broto et al. [46] (·· means aromatic).

values were 0.43 log unit lower than the calculated ones. The high standard deviation (0.83) was attributed to the variable strength of the hydrogen bonds.

Method of Ghose, Pritchett, and Crippen   The method of Ghose et al. is an atom contribution method where contributions have been evaluated for H and C atoms and heteroatoms (O, S, Se, N, P, F, Cl, Br, I) [48–50]. H-atom classification depends on the hybridization state, the oxidation number, and the heteroatom environment of the C atom to which the considered H atom is attached. Figure 13.4.4 illustrates the specification of H atoms in 4,4-difluorobutanoic acid.

Carbon atoms are classified depending on their hybridization and whether their neighbors are carbon atoms or heteroatoms. Halogen atoms are classified by the hybridization and oxidation state of the C atom to which they are attached. O, S, Se, N, and P are classified in different ways. The model uses 120 different atom-type descriptions and has been developed with a training set of 893 compounds. Observed versus calculated $\log K_{ow}$ showed a correlation coefficient of 0.926 and a standard deviation of 0.496. This method has been implemented in the Toolkit. Applications are shown in Figures 13.4.5 and 13.4.6 for the same compounds used to illustrate the Broto et al. method (Figs. 13.4.2 and 13.4.3).

Method of Suzuki and Kudo   The Suzuki and Kudo method [51] is based on a set of 1465 training compounds and 415 groups. Groups consist of a key group such as CH_2, CHBr, CF_3, SO_2, or ONO_2. To derive a specific group notation, a key group is further defined by its structural environment. For example, the carbonyl group, CO,

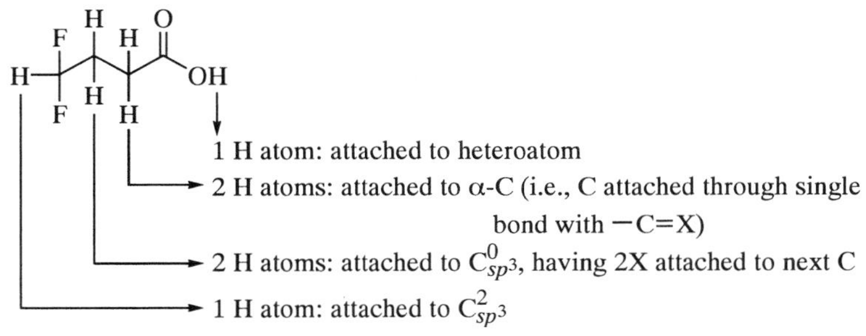

Figure 13.4.4 Classification of H atoms in 4,4-difluorobutanoic acid as defined in the model of Ghose et al. [49,50]. The subscript represents hybridization and the superscript is the formal oxidation number. X represents any heteroatom (O, N, S, P, Se, and halogens).

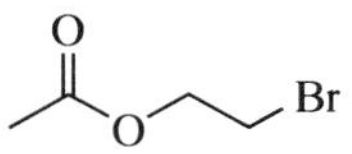

2-Bromoethyl ethanoate

Atomic group	Location	Contribution terms	
:CH$_3$R (No. 1)	ethanoate group	1($-$0.6771)	$-$0.6771
H [α-C] (No. 51)	ethanoate group	3(0.2099)	0.6297
:R$-$C(=X)$-$X (No. 40)	ethanoate group	1(0.0709)	0.0709
O= (No. 58)	ethanoate group	1($-$0.3514)	$-$0.3514
R$-$O$-$C=X (No. 60)	ethanoate group	1(0.2712)	0.2712
:CH$_2$RX (No. 6)	ethyl group	2($-$0.8370)	$-$1.6740
H [C^1_{sp3}] (No. 47)	ethyl group	4(0.3343)	1.3372
Br$-$C$-$ [C^1_{sp3}] (No. 91)	bromo group	1(1.0242)	1.0242

$$\log_{10} K_{ow} = 0.631$$
$$K_{ow} = 4.27$$

Figure 13.4.5 Estimation of K_{ow} for 2-bromoethyl ethanoate using the method of Ghose and Crippen [48] (constants from [50]).

in methyl phenyl ketone gets the notation CO$-$(C$_{ar}$)(C), where (C$_{ar}$) is an aromatic (sp^2) carbon and (C) is a single-bonded sp^3–carbon. The table of the 415 groups, their frequencies of use, and their contributions are provided by Suzuki and Kudo [51], complemented by a table of contribution values for extended groups such as polycyclic and heterocyclic substructures. The method has been implemented as CHEMICALC (Combined Handling of Estimation Methods Intended for Completely Automated Log P Calculation) and upgraded to an extended version, CHEMICALC2 [52]. CHEMICALC has been tested with 1686 compounds, including the 1465 training compounds. For the 221 test compounds, a correlation coefficient of 0.938 and an average absolute error, $\sum |\log K_{ow}(\text{obsd}) - \log K_{ow}(\text{est})|/221$, of 0.49 has been reported.

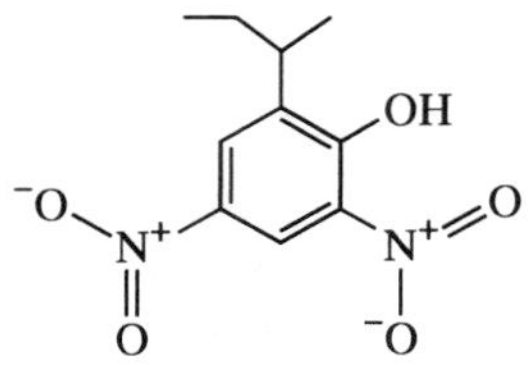

2,4-Dinitro-6-*sec*-butylphenol

Atomic group	Location	Contribution terms	
:Ar–NO$_2$ (No. 76)	nitro group	2(− 2.7640)	− 5.5280
:–O (No. 61)	nitro group	4(1.5810)	6.3240
O in phenol (No. 57)	hydroxy group	1(0.4860)	0.4860
H[heteroatom] (No. 50)	hydroxy group	1(− 0.3260)	− 0.3260
:CH$_3$R (No. 1)	*sec*-butyl group	2(− 0.6771)	− 1.3542
H [C$^0_{sp^3}$, 0X] (No. 46)	*sec*-butyl group	6(0.4418)	2.6508
:CH$_2$R$_2$ (No. 2)	*sec*-butyl group	1(− 0.4873)	− 0.4873
H [C$^0_{sp^3}$, 0X] (No. 46)	*sec*-butyl group	2(0.4418)	0.8836
:CHR$_3$ (No. 4)	*sec*-butyl group	1(− 0.3633)	− 0.3633
H [C$^0_{sp^3}$, 0X] (No. 46)	*sec*-butyl group	1(0.4418)	0.4418
:R–CH–R (No. 24)	benzene ring	2(0.0068)	0.0136
H [C$^0_{sp^2}$] (No. 47)	benzene ring	2(0.3343)	0.6686
:R–CR–R (No. 25)	benzene ring	1(0.1600)	0.1600
:R–CX–R (No. 26)	benzene ring	3(− 0.1033)	− 0.3099

$$\log_{10} K_{ow} = 3.259$$
$$K_{ow} = 1.82 \times 10^3$$

Figure 13.4.6 Estimation of K_{ow} for 2,4-dinitro-6-*sec*-butylphenol using the method of Ghose and Crippen [48] (constants from [50]).

Method of Nys and Rekker   The Nys and Rekker method [53,54] has been developed for mono- and di-substituted benzenes. The substituents considered are halogen atoms and hydroxyl, ether, amino, nitro, and carboxyl groups, for which contributions have been calculated by multiple regression analysis ($s = 0.106$, $r = 0.994$, $F = 1405$). Rekker discusses the extension of his approach to other compound classes, such as PAHs, pyridines, quinolines, and isoquinolines.

Method of Hansch and Leo   The Hansch and Leo method [5] differs from the methods above in its use of both (1) group contribution and (2) various factors accounting for certain structural features, such as unsaturation, branching, chains, rings, ring substitution pattern, multiple halogenation, proximity factors for polar groups, and intramolecular H-bond factors. The groups are either atom or polyatomic groups. Over 200 different group contribution values and 14 different factors are available. Lyman provides a detailed description of the method, its validity and illustrative examples on the application of this method [55]. Application of this method, for example, has been demonstrated with phenylalkylamides [56], where excellent agreement between estimated and newly measured K_{ow} have been found.

Further comparisons of experimental versus estimated values are discussed, for example, for hydroxyureas [57]. Based on new experimental K_{ow} for substituted α,N-diphenylnitrones and benzonitrile N-oxides, Kirchner et al. [58] evaluated contributions for several N-oxide groups for which contribution values were so far missing in the scheme of Hansch and Leo. Similarly, Finizio et al. [35] derived new contributions for three s-triazine groups. The groups and their contributions are shown in Figures 1.7.6 and 1.7.7.

CLOGP is a modified, computerized version of this method [59]. Chou and Jurs [60] describe its implementation and compare CLOGP estimation results with experimental values and with results derived using the method of Hopfinger and Battershell and the π substituent method (see below). Viswanadhan et al. [61] compare CLOGP estimation results for nucleosides and nucleoside bases with results derived using two other methods, a molecular orbital method and the method of Ghose and Crippen. They conclude that none of these methods take into account conformational flexibility or intramolecular hydrogen bonding, which can cause substantial discrepancy between observed and predicted K_{ow}. Limitations in using the method of Hansch and Leo and CLOGP have also been discussed for paracyclophane [62], for polyoxyethylene compounds [63], and for β-D-glucopyranosides [3]. However, these are not limitations particular to CLOGP but rather, limitations inherent to the group contribution approach in general.

Method of Hopfinger and Battershell The Hopfinger and Battershell method [64] is based on the solvent-dependent conformational anaylsis procedure (SCAP) to calculate the free energies of the solute at 300 K in water, F_w, and in octanol, F_o. The latter parameters are related to K_{ow} by the following relation:

$$\log_{10} K_{ow} = -0.735(F_w - F_o) \qquad (13.4.1)$$

F_w and F_o are calculated by SCAP from group contributions. Four different contributions, called the *hydration shell parameters*, have to be considered for each group and each solvent. They have been listed for the methyl, methylene, methine, vinyl, acetylene, aromatic C, halogens (except I), and various O-, S-, and N-containing groups [64].

Method of Camilleri, Watts, and Boraston The Camilleri et al. method [62] is based on an atom contribution derived as a solvent-accessible surface area of the individual atoms in a molecule. The authors have computed the surface areas for over 200 benzenes and PAHs substituted with one or more groups of the following type: alkyl, chloroalkyl, alkoxy, hydroxyl, amino, and carbonyl. Estimation results with this method have been compared with those obtained with CLOGP.

Method of Klopman and Wang The Klopman and Wang method [65] is based on the computer-aided structure evaluation (CASE) approach. Rather than including all groups in the estimation scheme, this method employs only groups that are identified by stepwise multiple regression analysis as the most significant groups contributing to K_{ow}. With a training set of 935 compounds, a relationship between K_{ow} and 39 contributions for atom-centered groups has been derived [$s = 0.39$,

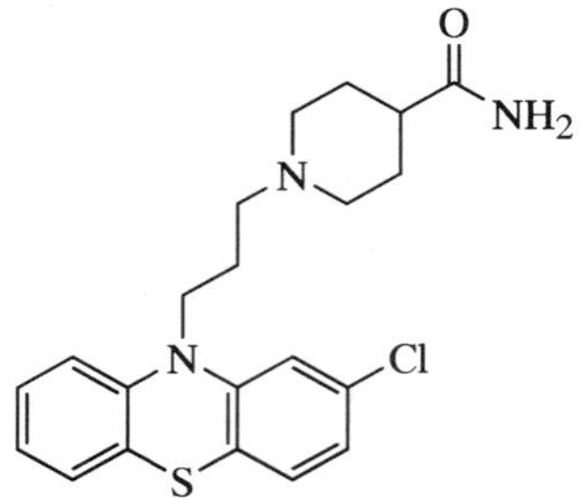

Figure 13.4.7   Pipamazine and its experimental value K_{ow} [65].

$r^2 = 0.93$, $F(39,895,005) = 316.5$]. The model has been reported to produce accurate estimations for complex molecules. Experimental and estimated K_{ow} values are shown for a validation set of various multifunctional compounds. Model application has been demonstrated by the authors for pipamazine (Figure 13.4.7), for which the estimated log K_{ow} value is 4.69, comparing favorably with the experimental value of 4.44.

Method of Klopman, Li, Wang, and Dimayuga   The Klopman et al. method [66] combines the "basic" group contribution approach with the computer aided structure evaluation (CASE) approach. The model equation is

$$\log_{10}K_{ow} = a + \sum b_i B_i + \sum c_j C_j \qquad (13.4.2)$$

where a, b_i, and c_j are regression coefficients, B_i is the number of occurrences of the ith basic group, and C_j is the number of occurrences of the jth correction factor identified by the CASE procedure. Basic groups are of two types: (1) atomic groups (C, F, Cl, Br, I), including specification of their hybridization and the number of H atoms attached, and (2) functional groups containing at least one heteroatom (N, O, P, S). The correction factors are additional contributions for specific substructures containing more than two nonhydrogen atoms. Overall, the model includes 98 different contributions derived from a set of 1663 compounds with diverse structures. A basic model (all $c_j = 0$) with 68 contributions had been derived first. Then substructures generated with the CASE program were labeled as active or inactive depending on whether they originated from compounds with a positive or a negative error determined as the difference between observed and retrofit values. Substructures were identified that could be classified as belonging to either the active or inactive class. These substructures were included to derive the final model according to eq. 13.4.2 [$n = 1663$, SD $= 0.3817$, $r^2 = 0.928$, $F(94,1568) = 217.77$]. The model has been tested by cross-validation, demonstrating the model's capability of predicting K_{ow} for simple as well as complex compounds.

13.5   SIMILARITY-BASED APPLICATION OF GCMs

π-**Substituent Constant**   The substituent constant, π_X, represents the difference between the log K_{ow} values of two compounds R–H and R–X [5]. In terms of the group contribution approach, π_X is the log K_{ow} difference resulting from the replacement of a hydrogen atom by a group X. Once the π constant has been derived from a given set of parent–congener pairs, R–H/R–X, the K_{ow} value of other congeners can be estimated using the following relationship:

$$(\log K_{ow})_{RX} = (\log K_{ow})_{RH} + \pi_X \qquad (13.5.1)$$

where π_X values for various functional groups have been evaluated [54]. The molecules of RX and RH are structurally similar with respect to their subgraph R, which is common to both molecules. Most π_X have been derived with R being an aromatic ring system. R may be an unsubstituted or a partly substituted ring system. In the latter case, however, π_X alters depending on the types of the substituents included in R and on the particular substitution pattern. Nakagawa et al. [67] conducted a detailed study to account for substitution patterns in multisubstituted benzenes. The model can be presented as follows:

$$(\log K_{ow})_{PhX_1,X_2,X_3,X_4,X_5,X_6} = (\log K_{ow})_{PhH} + \Delta \log K_{ow}\left(\sum X_c\right) \qquad (13.5.2)$$

where $PhX_1,X_2,X_3,X_4,X_5,X_6$ is a multisubstituted benzene and $\Delta \log K_{ow}$ is the overall substituent constant accounting for all replacements of H atoms by substituents X_i. For monosubstituted benzenes, $\Delta \log K_{ow}$ in eq. 13.5.2 is equivalent with π_X. For multisubstituted benzenes, $\Delta \log K_{ow}(\sum X_c)$ is formulated as a function of various substituent parameters accounting for stereoelectronic effects of the substituents X_i [67]. A set of 215 multisubstituted benzenes, including acetanilides, benzamides, nitrobenzenes, and anisols, has been used in evaluation of this approach.

Group Interchange Method of Drefahl and Reinhard   In the group interchange method (GIM) of Drefahl and Reinhard [68], the approach is not restricted to the replacement of H atoms. Group-by-group replacement (including insertion or deletion of bivalent groups) is allowed. The replacement of a terminal (monovalent) group G_x of molecule R–G_x by a different monovalent group G_y yields R–G_y. For an interchange of the terminal group G_x by G_y we write :

$$(\log K_{ow})_{R-G_y} = (\log K_{ow})_{R-G_x} + \Delta(\text{replacement of } G_x \text{ by } G_y) \qquad (13.5.3a)$$

For the interchange of the bivalent group G'_x by G'_y we write:

$$(\log K_{ow})_{R-G'_y-R'} = (\log K_{ow})_{R-G'_x-R'} + \Delta(\text{replacement of } G'_x \text{ by } G'_y) \qquad (13.5.3b)$$

Δ values depend on the interchanged groups (G_x, G_y or G'_x, G'_y) and on their neighbor groups. The derivation of Δ values will be demonstrated for the replacement of a methyl group, $-CH_3$, by a bromomethyl group, $-CH_2Br$. Table 13.5.1 lists log K_{ow} and Δ of corresponding compound pairs related by this particular structural difference. Fair agreement between the Δ values is observed. The lowest Δ is -0.25, the highest Δ is -0.31, and the mean Δ is -0.288. In all cases, R is an unbranched alkyl chain that builds the site's structural environment of interchangeable groups. In Figure 13.5.1 the evaluation of Δ is shown applying the GCM of Broto et al., of Ghose and Crippen, and of Hansch and Leo. The value derived with the latter method is closest to the value derived directly with experimental data in Table 13.5.1. In an analogous fashion, the derivation of Δ values for the replacement of a methyl group, $-CH_3$, by an ethyl group, $-CH_2CH_3$ is shown in Table 13.5.2 and Figure 13.5.2.

Drefahl and Reinhard [68] have designed an unambiguous notation system aimed to a compact, unique description of Δ (see also Section 1.7). Based on these notations for Δ, automatic estimation of K_{ow} is possible for compounds for which structurally similar compounds with known K_{ow} are available in database. This approach has been implemented in the Toolkit using the database of 600 compounds from the compilation of Sangster [1] as candidates. The GIM approach is demonstrated for 1-bromooctane in Figure 13.5.3. The recommended log K_{ow} value is 4.89 ± 0.35 [1].

---

1. Method of Broto et al. [46]:

Delete:	C–	0.631
Insert:	–C–	0.456
	Br–C–	0.620

$\Delta = -0.631 + 0.456 + 0.620 = 0.445$

2. Method of Ghose and Crippen [48, 50]:
Assume that methyl group is attached to methylene group

Delete:	:CH$_3$R	1(-0.6771)	-0.6771
	H at $C^0_{sp^3}$, no X attached to next C	3(0.4418)	1.3254
	:CH$_2$R$_2$	1(-0.4873)	-0.4873
	H at $C^0_{sp^3}$, no X attached to next C	2(0.4418)	0.8836
Insert:	Br at $C^1_{sp^3}$,	1(1.0242)	1.0242
	CH$_2$RX	1(-0.8370)	-0.8370
	H at $C^1_{sp^3}$	2(0.3343)	0.6686
	:CH$_2$R$_2$	1(-0.4873)	-0.4873
	H at $C^0_{sp^3}$, 1 X attached to next C	2(0.3695)	0.7390

$$\Delta = -(-0.6771 + 1.3254 - 0.4873 + 0.8836) + (1.0242 - 0.8370 + 0.6686$$
$$-0.4873 + 0.7390) = -1.0446 + 1.1075 = 0.0629$$

3. Method of Hansch and Leo [5]:

Delete:	–H	0.23
Insert:	–Br	0.20
	F$_b$	-0.12

$\Delta = -0.23 + (0.20 - 0.12) = -0.15$

Figure 13.5.1   Derivation of Δ (replacement of $-CH_3$ by $-CH_2Br$) using GCMs.

TABLE 13.5.1 Derivation of Δ (replacement of $-CH_3$ by $-CH_2Br$) Using Experimental log K_{ow}

Target Compound		Source Compound		
Name	log K_{ow}^a	Name	log K_{ow}^a	Δ
1-Bromopentane	3.37	n-Pentane	3.62	-0.25
1-Bromohexane	3.80	n-Hexane	4.11	-0.31
1-Bromoheptane	4.36	n-Heptane	4.66	-0.30
1-Bromooctane	4.89	n-Octane	5.18	-0.29
			$\Delta_{mean} =$	-0.28_8

^{a} Data from [13].

TABLE 13.5.2 Derivation of Δ (replacement of $-CH_3$ by $-CH_2CH_3$) Using Experimental log K_{ow}

Target Compound		Source Compound		
Name	log K_{ow}^a	Name	log K_{ow}^a	Δ
n-Hexane	4.11	n-Pentane	3.62	0.49
1-Hexyne	2.73	1-Pentyne	2.12	0.61
n-Hexylbenzene	5.52	n-Pentylbenzene	4.90	0.62
1-Bromohexane	3.80	1-Bromopentane	3.37	0.43
1-Hexanol	1.84	1-Pentanol	1.34	0.50
			$\Delta_{mean} = 0.53_0$	

^{a} Data from [13].

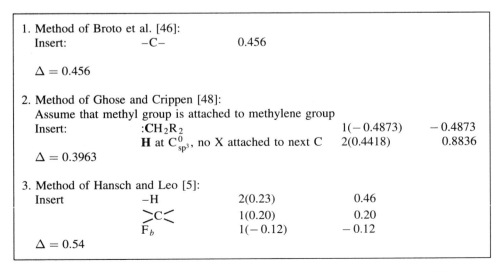

Figure box:

1. Method of Broto et al. [46]:
   Insert: $-C-$ 0.456

   $\Delta = 0.456$

2. Method of Ghose and Crippen [48]:
   Assume that methyl group is attached to methylene group
   Insert: $:CH_2R_2$ $1(-0.4873)$ -0.4873
            H at $C_{sp^3}^0$, no X attached to next C $2(0.4418)$ 0.8836
   $\Delta = 0.3963$

3. Method of Hansch and Leo [5]:
   Insert $-H$ $2(0.23)$ 0.46
            $>C<$ $1(0.20)$ 0.20
            F_b $1(-0.12)$ -0.12
   $\Delta = 0.54$

Figure 13.5.2 Derivation of Δ (replacement of $-CH_3$ by $-CH_2Br$) using GCMs.

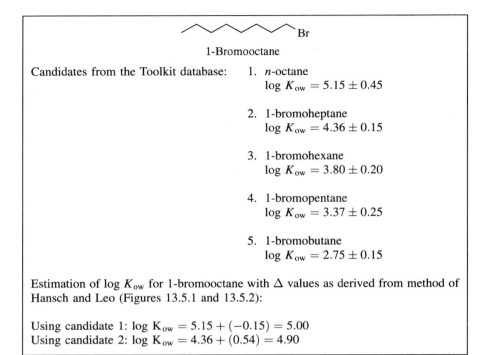

1-Bromooctane

Candidates from the Toolkit database:
1. *n*-octane
   $\log K_{ow} = 5.15 \pm 0.45$

2. 1-bromoheptane
   $\log K_{ow} = 4.36 \pm 0.15$

3. 1-bromohexane
   $\log K_{ow} = 3.80 \pm 0.20$

4. 1-bromopentane
   $\log K_{ow} = 3.37 \pm 0.25$

5. 1-bromobutane
   $\log K_{ow} = 2.75 \pm 0.15$

Estimation of $\log K_{ow}$ for 1-bromooctane with Δ values as derived from method of Hansch and Leo (Figures 13.5.1 and 13.5.2):

Using candidate 1: $\log K_{ow} = 5.15 + (-0.15) = 5.00$
Using candidate 2: $\log K_{ow} = 4.36 + (0.54) = 4.90$

Figure 13.5.3 Estimation of K_{ow} for 1-bromooctane using GIM.

$$R \diagdown_O \diagup \diagdown_{O} \diagdown_{H} {}_{n}$$

Figure 13.5.4 Polyoxyethylene alkylphenyl and alkyl ethers (R = alkyl/alkylphenyl, n = number of $-CH_2-CH_2-O-$groups).

GIM of Schüürmann for Oxyethylated Surfactants Schüürmann [63] studied K_{ow} and its relation to aquatic toxicity for polyoxyethylene compounds of the type shown in Figure 13.5.4. The contribution of the $-CH_2-CH_2-O-$ group is not a constant but depends on the length of the oxyethylene chain. For the insertion of a $-CH_2-CH_2-O-$ group, Schüürmann reports a variation from -0.10 for long-chain molecules to -0.19 for short-chain molecules.

REFERENCES

1. Sangster, J., Octanol–Water Partition Coefficients of Simple Organic Compounds. *J. Phys. Chem. Ref. Data*, 1989: **18**, 1111–1229.

2. Lyman, W. J., Octanol/Water Partition Coefficient, in *Handbook of Chemical Property Estimation Methods*, W. J. Lyman, W. F. Reehl, and D. H. Rosenblatt, Editors, 1990. Washington, DC: American Chemical Society.

3. Kim, K. H., and Y. C. Martin, Comparison of Calculated Versus Measured Partition Coefficients of Some Phenyl. β-D-Glucopyranosides. *J. Pharm. Sci.*, 1986: **75**, 637–638.

4. Doucette, W. J., and A. W. Andren, Estimation of Octanol / Water Partition Coefficients: Evaluation of Six Methods for Highly Hydrophobic Aromatic Hydrocarbons. *Chemosphere*, 1988: **17**, 345–359.

5. Hansch, C., and A. Leo, *Substituent Constants for Correlation Analysis in Chemistry and Biology*, 1979. New York: Wiley.

6. Chessells, M., D. W. Hawker, and D. W. Connell, Influence of Solubility in Lipid on Bioconcentration of Hydrophobic Compounds. *Ecotoxicol. Environ. Safety*, 1992: **23**, 260–273.

7. Tanford, C., The Hydrophobic Effect: Formation of Micelles and Biological Membranes, 1973. New York: Wiley.

8. van Haelst, A. G., et al., Determination of *n*-Octanol / Water Partition Coefficients of Tetrachlorobenzyltoluenes Individually and in a Mixture by the Slow Stirring Method. *Chemosphere*, 1994: **29**, 1651–1660.

9. De Bruijn, J., et al., Determination of Octanol / Water Partition Coefficients for Hydrophobic Organic Chemicals with the "Slow-Stirring" Method. *Environ. Toxicol. Chem.*, 1989: **8**, 499–512.

10. Zachara, J. M., et al., Sorption of Binary Mixtures of Aromatic Nitrogen Heterocyclic Compounds on Subsurface Materials. *Environ. Sci. Technol.*, 1987: **21**, 397–402.

11. Isnard, P., and S. Lambert, Aqueous Solubility and *n*-Octanol / Water Partition Coefficient Correlations. *Chemosphere*, 1989: **18**, 1837–1853.

12. Bowman, B. T., and W. W. Sans, Determination of Octanol–Water Partitioning Coefficients (KOW) of 61 Organophosphorus and Carbamate Insecticides and Their Relationship to Respective Water Solubility (S) Values. *J. Environ. Sci. Health*, 1983: **B18**(6), 667–683.

13. Tewari, Y. B., et al., Aqueous Solubility and Octanol / Water Partition Coefficient of Organic Compounds at 25.0°C. *J. Chem. Eng. Data*, 1982: **27**, 451–454.

14. Arbuckle, W. B., Estimating Activity Coefficients for Use in Calculating Environmental Parameters. *Environ. Sci. Technol.*, 1983: **17**, 537–542.

15. Chen, F., J. Holten-Andersen, and H. Tyle, New Developments of the UNIFAC Model for Environmental Application. *Chemosphere*, 1993: **26**, 1325–1354.

16. Collander, R., The Partitioning of Organic Compounds Between Higher Alcohols and Water. *Acta Chem. Scand.*, 1951: **5**, 774–780.

17. Muller, N., When Is a Trifluoromethyl Group More Lipophilic Than a Methyl Group: Partition Coefficients and Selected Chemical Shifts of Aliphatic Alcohols and Trifluoroalcohols. *J. Pharm. Sci.*, 1986: **75**, 987–991.

18. He, Y., et al., Determination and Estimation of Physicochemical Properties for Phenylsulfonyl Acetates. *Chemosphere*, 1995: **30**, 117–125.

19. Eadsforth, C. V., and P. Moser, Assessment of Reverse-Phase Chromatographic Methods for Determining Partition Coefficients. *Chemosphere*, 1983: **12**, 1459–1475.

20. Veith, G. D., N. M. Austin, and R. T. Morris, A Rapid Method for Estimating log P for Organic Chemicals. *Water Res.*, 1979: **13**, 43–47.

21. Chin, Y.-P., W. J. Weber, Jr., and T. C. Voice, Determination of Partition Coefficients and Aqueous Solubilities by Reverse Phase Chromatography: II. *Water Res.*, 1986: **20**, 1443–1450.

22. Burkhard, L. P., D. W. Kuehl, and G. D. Veith, Evaluation of Reverse Phase Liquid Chromatography / Mass-Spectrometry for Estimation of *N*-Octanol / Water Partition Coefficients. *Chemosphere*, 1985: **14**, 1551–1560.

23. McDuffie, B., Estimation of Octanol/Water Partition Coefficients for Organic Pollutants Using Reverse- Phase HPLC. *Chemosphere*, 1981: **10**, 73–83.

24. Rapaport, R. A., and S. J. Eisenreich, Chromatographic Determination of Octanol–Water Partition Coefficients (K_{ow}'s) for 58 Polychlorinated Biphenyl Congeners. *Environ. Sci. Technol.*, 1984: **18**, 163–170.

25. Karcher, W., *Spectral Atlas of Polycyclic Aromatic Compounds Including Data on Occurrence and Biological Activity*, 1985. Dordrecht, The Netherlands: D. Reidel.

26. Sarna, L. P., P. E. Hodge, and G. R. B. Webster, Octanol–Water Partition Coefficients of Chlorinated Dioxins and Dibenzofurans by Reversed-Phase HPLC Using Several C18 Columns. *Chemosphere*, 1984: **13**, 975–983.

27. Burkhard, L. P., and D. W. Kuehl, *N*-Octanol/Water Partition Coefficients by Reverse Phase Liquid Chromatography/Mass Spectrometry for Eight Tetrachlorinated Planar Molecules. *Chemosphere*, 1986: **15**, 163–167.

28. Biagi, G. L., et al., R_m Values of Phenols: Their Relationship with Log *P* Values and Activity. *J. Med. Chem.*, 1975: **18**, 868–873.

29. Budvári-Bárány, Z., et al., Investigation of Structure–Property Relationships for Imida-zoquinolone Derivatives. I. Relation Between Partition Coefficient and Chromatographic Retention. *J. Liq. Chromatogr.*, 1990: **13**, 1485–1497.

30. Takács-Novák, K., et al., Relationship Study Between Reversed Phase HPLC Retention and Octanol/Water Partition Among Amphoteric Compounds. *J. Liquid Chromatogr.*, 1995: **18**, 807–825.

31. Niemi, G. J., et al., Prediction of Octanol/Water Partition Coefficient (K_{ow}) with Algorithmically Derived Variables. *Environ. Toxicol. Chem.*, 1992: **11**, 893–900.

32. De Bruijn, J., et al., Determination of Octanol/Water Partition Coefficients for Hydro-phobic Organic Chemicals with the "Slow-Stirring" Method. *Environ. Toxicol. Chem.*, 1989: **8**, 499–512.

33. Shiu, W. Y., et al., Physical-Chemical Properties of Chlorinated Dibenzo-*p*-dioxins. *Environ. Sci. Technol.*, 1988: **22**, 651–658.

34. Kier, L. B., and L. H. Hall, *Molecular Connectivity in Chemistry and Drug Research*, 1976. San Diego, CA: Academic Press.

35. Finizio, A., et al., Different Approaches for the Evaluation of K_{ow} for *s*-Triazine Herbicides. *Chemosphere*, 1991: **23**, 801–812.

36. Govers, H., et al., Experimental Determination and Prediction of Partition Coefficients of Thioureas and Their Toxicity to *Photobacterium phosphoreum*. *Chemosphere*, 1986: **22**, 383–393.

37. Basak, S. C., G. J. Niemi, and G. D. Veith, Optimal Characterization of Structure for Prediction of Properties. *J. Math. Chem.*, 1990: **4**, 185–205.

38. Saçan, M. T., and Y. Inel, Application of the Characteristic Root Index Model to the Estimation of *N*-Octanol/Water Partition Coefficients: Polychlorinated Biphenyls. *Chemosphere*, 1995: **30**, 39–50.

39. Yang, Y.-Q., L. Xu, and C.-Y. Hu, Extended Adjacency Matrix Indices and Their Applications. *J. Chem. Inf. Comput. Sci.*, 1994: **34**, 1140–1145.

40. Moriguchi, I., Y. Kanada, and K. Komatsu, Van der Waals Volume and the Related Parameters for Hydrophobicity in Structure–Activity Studies. *Chem. Pharm. Bull.*, 1976: **24**, 1799–1806.

41. Leo, A., C. Hansch, and Y. C. Jow, Dependence of Hydrophobocity of Apolar Molecules on Their Molecular Volume. *J. Med. Chem.*, 1976: **19**, 611–615.

42. Lewis, D. F. V., The Calculation of Molar Polarizabilities by the CNDO/2 Method: Correlation with the Hydrophobic Parameter, Log *P*. *J. Comput. Chem*, 1989: **10**, 145–151.

43. Bodor, N., Z. Gabanyi, and C.-K. Wong, A New Method for the Estimation of Partition Coefficient. *J. Am. Chem. Soc.*, 1989: **111**, 3783–3786.

44. Bodor, N., M.-J. Huang, and A. Harget, Neural Network Studies: 3. Prediction of Partition Coefficients. *J. Mol. Struct. (Theochem.)*, 1994: **309**, 259–266.

45. Korenman, I. M., N. Y. Gurevich, and T. G. Kulagina, Extraction of Certain *n*-Alkylamines from Aqueous Solutions. *Zh. Prikl. Khim.* (J. Appl. Chem. USSR), 1973: **46EE**, 726–728.

46. Broto, P., G. Moreau, and C. Vandycke, Molecular Structure: Perception, Autocorrelation Descriptor and SAR Studies. System of Atomic Contributions for the Calculation of the *n*-Octanol / Water Partition Coefficients. *Eur. J. Med. Chem.-Chim. Ther.*, 1984: **19**, 71–78.

47. Convard, T., et al., SmilogP: A Program for a Fast Evaluation of Theoretical Log *P* from the Smiles Code of a Molecule. *Quant. Struct.-Act. Relat.*, 1994: **13**, 34–37.

48. Ghose, A. K., and G. M. Crippen, Atomic Physicochemical Parameters for Three Dimensional Structure Directed Quantitative Structure–Activity Relationships: I. *J. Comput. Chem.*, 1986: **7**, 565–577.

49. Ghose, A. K., A. Pritchett, and G. M. Crippen, Atomic Physicochemical Parameters for Three Dimensional Structure Directed Quantitative Structure–Activity Relationships: III. Modeling Hydrophobic Interactions. *J. Comput. Chem.*, 1988: **9**, 80–90.

50. Viswanadhan, V. N., et al., Atomic Physicochemical Parameters for Three Dimensional Structure Directed Quantitative Structure–Activity Relationships. IV. Additional Parameters for Hydrophobic and Dispersive Interaction and Their Application for an Automated Superposition of Certain Naturally Occurring Nucleoside Antibiotics. *J. Chem. Inf. Comput. Sci.*, 1989: **29**, 163–172.

51. Suzuki, T., and Y. Kudo, Automatic Log *P* Estimation Based on Combined Additive Modeling Methods. *J. Comput.-Aid. Mol. Des.*, 1990: **4**, 155–198.

52. Suzuki, T., Development of an Automatic Estimation System for Both the Partition Coefficient and Aqueous Solubility. *J. Comput.-Aid. Mol. Des.*, 1991: **5**, 149–166.

53. Nys, G. G., and R. F. Rekker, The Concept of Hydrophobic Fragmental Constants (*f*-Values): II. Extension of Its Applicability to the Calculation of Lipophilicities of Aromatic and Heteroaromatic Structures. *Eur. J. Med. Chem.-Chim. Ther.*, 1974: **9**, 361–375.

54. Rekker, R. F., A. M. ter Laak, and R. Mannhold, On the Reliability of Calculated Log *P*-values: Rekker, Hansch / Leo and Suzuki Approach. *Quant. Struct.-Act. Relat.*, 1993: **12**, 152–157.

55. Lyman, W. J., W. F. Reehl, and D. H. Rosenblatt, *Handbook of Chemical Property Estimation Methods*, 1990. Washington, DC: American Chemical Society.

56. Hernandez-Gallegos, Z. and P. A. Lehmann, Partition Coefficients of Three New Anticonvulsants. *J. Pharm. Sci.*, 1990: **79**, 1032–1033.

57. Parker, G. R., T. L. Lemke, and E. C. Moore, Effect of the Solvent-Dependent Conformational System of Hydroxyureas on Predicted vs. Observed Log *P*. *J. Med. Chem.*, 1977: **20**, 1221–1225.

58. Kirchner, J. J., et al., Octanol–Water Partition Coefficients of Substituted. α,N-Diphenylnitrones and Benzonitrile *N*-Oxides. *J. Pharm. Sci.*, 1985: **74**, 1129–1130.

59. Manual, S., *MedChem Manual. Release 3*, 1985. Claremont, CA: Pomona College.

60. Chou, J. T., and P. C. Jurs, Computer-Assisted Computation of Partition Coefficients from Molecular Structure Using Fragment Constants. *J. Chem. Inf. Comput. Sci.*, 1979: **19**, 172–178.

61. Viswanadhan, V. N., et al., Assessment of Methods Used for Predicting Lipophilicity: Application to Nucleosides and Nucleoside Bases. *J. Comput. Chem.*, 1992: **14**, 1019–1026.

62. Camilleri, P., S. A. Watts, and J. A. Boraston, Surface Area Approach to Determination of Partition Coefficients. *Chem. Soc. Perkin Trans.*, 1988: **II**, 1699ff.

63. Schüürmann, G., QSAR Analysis of the Acute Toxicity of Oxyethylated Surfactants. *Chemosphere*, 1990: **21**, 467–478.

64. Hopfinger, A. J., and R. D. Battershell, Application of SCAP to Drug Design: 1. Prediction of Octanol–Water Partition Coefficients Using Solvent-Dependent Conformational Analyses. *J. Med. Chem.*, 1976: **19**, 569–573.

65. Klopman, G., and S. Wang, A Computer Automated Structure Evaluation (CASE) Approach to Calculation of Partition Coefficient. *J. Comput. Chem.*, 1991: **12**, 1025–1032.

66. Klopman, G., et al., Computer Automated Log *P* Calculation Based on an Extended Group Contribution Approach. *J. Chem. Inf. Comput. Sci.*, 1994: **34**, 752–781.

67. Nakagawa, Y., et al., Analysis and Prediction of Hydrophobicity Parameters of Substituted Acetanilides, Benzamides and Related Aromatic Compounds. *Environ. Toxicol. Chem.*, 1992: **11**, 901–916.

68. Drefahl, A., and M. Reinhard, Similarity-Based Search and Evaluation of Environmentally Relevant Properties for Organic Compounds in Combination with the Group Contribution Approach. *J. Chem. Inf. Comput. Sci.*, 1993: **33**, 886–895.

CHAPTER 14

SOIL–WATER PARTITION COEFFICIENT

14.1 DEFINITION

The soil–water partition coefficient, $K_{\text{soil}-\text{water}}$ is a "conditional" and not a fundamental physicochemical compound property. $K_{\text{soil}-\text{water}}$ is included here because of its great practical significance. Its value depends on a number of soil and solution characteristics, such as the organic carbon (OC) or organic matter (OM) content, clay content and type, pore volume, pore size and distribution, and solution conditions. $K_{\text{soil}-\text{water}}$ can be defined as the ratio of the sorbate's mass sorbed per unit volume of soil to the mass dissolved per unit volume of aqueous phase with both phases at equilibrium:

$$K_{\text{soil}-\text{water}} = (m_{\text{soil}}/V_{\text{soil}})(m_{\text{water}}/V_{\text{water}}) \qquad (14.1.1)$$

The $K_{\text{soil}-\text{water}}$ partition coefficient is dimensionless. When determined in batch tests sorption is typically indicated as the soil–water distribution coefficeint, K_d, in which the soil is quantified in terms of mass rather than volume:

$$K_d = (m_{\text{soil}}/X_{\text{soil}})/C_e \qquad (14.1.2)$$

where X_{soil} indicates the mass of soil that is equilibrated with the sorbate solution and C_e is the aqueous phase equilibrium concentration. The units of K_d are in volume mass^{-1}, typically in $\text{L}\,\text{kg}^{-1}$. K_d can be obtained from $K_{\text{soil}-\text{water}}$ through the relationship

$$K_d = K_{\text{soil}-\text{water}}/\rho_s$$

where ρ_s is the average particle density in kg/L. The Freundlich equation is used to account for the concentration dependence of $m_{\text{soil}}/X_{\text{soil}}$ on C_e

$$m_{\text{soil}}/X_{\text{soil}} = K_p \cdot C_e^{1/N} \qquad (14.1.3)$$

where K_p and $1/N$ are the Freundlich sorption coefficient (in units of volume mass^{-1}) and the dimensionless Freundlich exponent respectively. The Freundlich exponent indicates nonlinearity and is typically less than one [1–4]. Often $1/N$ is sufficiently close to one so that K_d can be assumed to be constant for relatively narrow ranges in solution concentration. The organic carbon–water partioning coefficient, K_{oc}, is defined assuming that sorption by soil organic matter, X_{som} is independent of the mineral matrix:

$$K_{oc} = (m_{soil}/X_{som})/C_e \qquad (14.1.4)$$

The soil organic matter content is mostly determined as the organic carbon content X_{soc}. f_{oc} is defined as

$$f_{oc} = X_{soc}/X_{soil} \qquad (14.1.5)$$

where X_{soc} and X_{soil} are the organic carbon content and the total soil mass, respectively. The organic carbon content is often indicated in percent ($100 f_{oc} = \%OC$). If sorption by the mineral matrix contribution is insignificant, we can write for K_d

$$K_d = f_{oc} K_{oc} \qquad (14.1.6)$$

The estimation procedures discussed below apply to K_{oc}, $K_{soil-water}$ is obtained knowing f_{oc} and ρ_s. Sometimes, it is desirable to express sorption in terms of the organic matter content. K_{om} is related to K_{oc} by

$$K_{om} = K_{oc}/f_c \qquad (14.1.7)$$

where f_c indicates the organic carbon content of the organic matter. In a study with halogenated aliphatic compounds, K_{oc} and $1/N$ were found to depend on the age and origin of the organic matter [5]. K_{oc} estimations should be limited to soils with relatively high f_c since inorganic matrix effects are often significant and to intermediate concentrations ranges since deviation from isotherm linearity can lead to significant errors. It has been widely reported in the literature that K_{oc} values appear to increase with decreasing f_{oc} suggesting significant mineral contribution to sorption. Sorption and desorption rates may be extremely slow due to hindered diffusion in micro (nano-scale) pores as discussed by Luthy et al. [1]. Estimates for aquifer materials with low f_{oc} may yield erroneous results if sorption by the inorganic fraction [4] and slow diffusion are not considered [6].

Uses for $K_{soil-water}$

- To predict a compound's mobility in soil and aquifers
- To estimate a compound's tendency to accumulate in sediments
- To assess the bioavailability of a compound
- To estimate the tendency of a compound to evaporate from soil

---

Current models allow order-of-magnitude K_{oc} estimations but lack the fine-scale estimation capabilities that account for the variation of soil properties. The most frequently used estimation techniques used for K_{oc} are estimation by water solubility, octanol–water partition coefficient, reversed-phase high-pressure liquid chromatography (HPLC) capacity factor, and by molecular parameters, topological indices, and solvation energy relationships. Gawlik et al. compiled the available K_{oc} correlations for non-ionic compounds and found 24 equations for water solubility, 76 for octanol–water coefficient, 35 for RP–HPLC capacity factors, 24 for topological indices, 38 for molecular parameters, and 14 mixed or miscellaneous variables [7].

Temperature Dependence   Values of K_{oc} have usually been measured at temperatures between 20 and 25°C. Temperature caused changes in K_{oc} are expected to be similar to those of K_{ow}. Werth and Reinhard [8] studied the influence of temperature on TCE sorption by natural sediments soils, and aquifer material. In agreement with theoretical considerations, they found small heat effects under conditions when the soil organic matter was assumed to be the dominant sorbent phase.

pH Dependence   Polar compounds may ionize depending on the environmental pH. The degree of dissociation influences the sorption of these compounds. The nonionized forms of weak carbonic acids adsorb much stronger than the corresponding anions. Weak bases in the protonated form are adsorbed more strongly than in the unprotonated form [2]. The single solute sorption of basic compounds such as pyridine, quinoline, and acridine is higher in acidic soil, reflecting stronger sorption of the protonated form [9]. The pH dependence of chlorophenols has been studied by Schellenberg et al. [10], Lagas [11], and Lee et al. [12]. For 2,3,4,6-tetrachlorophenol, the log K_{oc} can vary from 2.59 for the dissociated form to 3.90 for the undissociated form [13]. Wang et al. [14] have investigated the binding capacity of atrazine and its hydrolysis product hydroxyatrazine as a function of pH and soil characteristics.

14.2   PROPERTY–SOIL WATER PARTITIONING RELATIONSHIPS

K_{oc} Estimation Using K_{ow}   The property used most often in estimation models for K_{oc} is K_{ow}. Correlations between K_{oc} and K_{ow} are represented by the following equation:

$$\log K_{oc} = a_0 + a_1 \log K_{ow} \qquad (14.2.1)$$

Gawlik et al. [7] and Lyman et al. [15] provide an overview of K_{oc} versus K_{ow} equations and Sabljic compares estimation results derived with various K_{ow}-based models [14]. Sabljic discusses and compares these models with respect to the inaccuracy and incompatibility of the experimental K_{ow} and K_{oc} data [16]. Most of those models apply for polynuclear aromatics, halogenated hydrocarbons, or certain classes of pesticides. Abdul et al. [17] report excellent agreement between the K_{oc} versus K_{ow} model derived with their experimental data and the model derived with a larger set of data. Paya-Perez et al. [18] have studied relationships for chlorobenzenes and PCBs. They found, however, that the corresponding correlation based on S_w as the independent variable is better than the one based on K_{ow}. Vowles and Mantoura

[19] compare the K_{oc} versus K_{ow} correlation for alkylbenzenes and alkylnaphthalenes with corresponding K_{oc} versus *HPLC capacity factor* correlations. A similar study has been reported by Hodson and Williams [20] for a more diverse set of compounds.

Bintein and Devillers [3] developed a model for alkylbenzenes, PAHs, halogenated alkanes, chlorinated benzenes, PCBs, and acids and bases. Their model has been derived from 229 K_p values (with $\%OC \geq 0.1$) recorded for 53 compounds. A test was performed on 500 other K_p values for 87 compounds. The model requires the input of the system parameter $\%OC$ and of two compound properties, K_{ow} and pK_a:

$$\log_{10}K_p = 0.25 + 0.93\log_{10}K_{ow} + 1.09\log_{10}\%OC + 0.32CFa - 0.55CFb$$
$$n = 229, \quad s = 0.433, \quad r = 0.966, \quad F = 786.07, p < 0.01\% \qquad (14.2.2)$$

where CFa is related to the anionic species concentration by

$$CFa = -\log_{10}(1 + 10^{pH-pKa}) \qquad (14.2.3a)$$

and CFb is related to the cationic species concentration by

$$CFb = -\log_{10}[1 + 10^{pKa-(pH-2)}] \qquad (14.2.3b)$$

If a compound is nonacid or nonbase, CFa and CFb must equal zero.

LSER Approach of He, Wang, Han, Zhao, Zhang, and Zou   The LSER approach has been described for aqueous solubility in Section 11.4. In analogy to eq. 13.2.8 for K_{ow}, He et al. [21] have derived the following relationship for phenylsulfonyl alkanoates:

$$\log_{10}K_{oc} = -(1.821 \pm 0.372) + \frac{(2.175 \pm 0.071)V_i}{100} - (0.666 \pm 0.102)\pi^*$$
$$- (1.260 \pm 0.182)\beta \qquad n = 28, \quad s = 0.067, \quad r = 0.988$$
$$(14.2.3)$$

where K_{oc} is at 25°C and V_i, π^* and β are as defined in 11.4 (Solvatochromic Approach).

14.3   STRUCTURE–SOIL WATER PARTITIONING RELATIONSHIPS

Model of Bahnick and Doucette   Molecular connectivity indices have been applied to establish structure–soil water partitioning relationships for various classes of compounds. Bahnick and Doucette [22] briefly review such models and present a new model for a variety of organic compounds, including halogenated alkanes, PAHs, chlorobenzenes, PCBs, and different pesticide classes. The model is

$$\log_{10}K_{oc} = 0.64 + 0.53(\pm0.04)\,^1\chi + 2.09(\pm0.22)\Delta^1\chi^\nu \qquad n = 56, \quad s = 0.34,$$
$$r = 0969, \quad F_{2,53} = 411 \qquad (14.3.1)$$

where

$$\Delta^1\chi^v = (^1\chi^v)_{np} - {}^1\chi^v \qquad (14.3.2)$$

In eq. 14.3.2, $^1\chi^v$ is the index for the heteroatom-containing molecule and $(^1\chi^v)_{np}$ is the index for the corresponding, nonpolar (*np*) hydrocarbon equivalent. Bahnick and Doucette [20] demonstrate the calculation of these descriptors for 2-chloroacetanilide. $\Delta^1\chi^v$ accounts for nondispersive molecular interaction. Testing this model on a validation set of 40 structurally diverse compounds resulted in a standard deviation for the experimental versus estimated values of 0.37. The comparison of this value with the standard error of estimate ($s = 0.34$) from the regression model suggests this model can be used confidently within the range of these structures.

14.4 GROUP CONTRIBUTION APPROACHES FOR SOIL–WATER PARTITIONING

Model of Okouchi and Saegusa For hydrocarbons and halogenated hydrocarbons, the following model has been suggested [23]:

$$\log_{10} K_{om} = 0.16 + 0.62\text{AI} \qquad n = 72, \quad s = 0.341, \quad r = 0967 \qquad (14.4.1)$$

with AI being the adsorbability index calculated as follows:

$$\text{AI} = \sum A + \sum I \qquad (14.4.2)$$

where A and I represent the atomic and group contributions, respectively, summed over all contributions present in the molecule. The atomic and group contributions are shown in Table 14.4.1.

Evaluating their test set, Okouchi and Saegusa concluded that the contribution of the I index could be ignored.

TABLE 14.4.1   Contributions to the Adsorbability Defined in Eq. 14.4.2

Group	A	Group	
C	0.26	Aliphatic	
H	0.12	−OH (alcohols)	− 0.53
N	0.26	−O− (ethers)	− 0.36
O	0.17	−CHO (aldehydes)	− 0.25
S	0.54	N (amines)	− 0.58
Cl	0.59	−COOR (ester)	− 0.28
Br	0.86	＞C=O (ketones)	− 0.30
NO$_2$	0.21	−COOH (fatty acids)	− 0.03
−C=C−	0.19	Aromatics	
iso	− 0.12	−OH, −O−, N, −COOR	
tert.	− 0.32	＞C=O, −COOH	0
cyclo	− 0.28	α-Amino acids	− 0.155

Source: Adapted from Ref. [23].

Model of Meylan et al. Meylan et al. [13] have presented an estimation model for K_{oc} that combines group additivity with molecular connectivity correlation. Their model is based on a linear nonpolar model relating K_{oc} to a molecular connectivity index and a polar model derived from the former model by introducing additional polarity correction factors, P_f, for 26 N-, O-, P-, and S-containing groups. The nonpolar model applies to halogenated and nonhalogenated aromatics, PAHs, halogenated aliphatics, and phenols:

$$\log_{10} K_{oc} = 0.62 + 0.53\,{}^1\chi \qquad n = 64, \quad s = 0.267, \quad r = 0.978, \quad F = 1371$$

$$(14.4.3a)$$

The polar model applies to all other compounds for which the fragments can be linked to a corrective contribution in the P_f list:

$$\log_{10} K_{oc} = 0.62 + 0.53\,{}^1\chi + \sum P_f N \qquad (14.4.3b)$$

where the products summation is over all applicable P_f factors for polar groups multiplied by the number of times that the group occurs in the molecule, N, except for certain fragments that are counted only once. A detailed description of the group specification, statistical performance, and model validation is available in the original source [13]. Meylan and Howard have developed the program PCKOC [24] to estimate K_{oc} from SMILES input. This model has also been incorporated into the Toolkit.

Lohninger [25] has taken the approach of Meylan et al. [13] one step further by combining group contributions and topological indices. Lohninger applied a set of 201 pesticides containing only the elements C, H, O, N, S, P, F, Cl, and Br to compare a multiple linear regression model and a neural network (radial basis function network) model. His final model is a linear regression model with two molecular descriptors (${}^0\chi^v$ and $n34$) and nine group contributions as independent variables [$n = 120$, $s = 0.5559$ (log K_{oc} units), $r = 0.8790$, $F = 33.377$].

REFERENCES

1. Luthy, R. G., et al., Sequestration of Hydrophobic Organic Contaminants by Geosorbents. *Environ. Sci. Technol.*, 1997: **31**(12), 3341–3347.

2. von Oepen, B., W. Kördel, and W. Klein, Sorption of Nonpolar and Polar Compounds to Soils: Process, Measurement and Experience with the Applicability of the Modified OECD-Guideline 106. *Chemosphere*, 1991: **22**, 285–304.

3. Bintein, S., and J. Devillers, QSAR for Organic Chemical Sorption in Soils and Sediments. *Chemosphere*, 1994: **28**, 1171–1188.

4. Farrell, J., and M. Reinhard, Desorption of Halogenated Organics from Model Solids, Sediments, and Soil Under Unsaturated Conditions: 1. Isotherms. *Environ. Sci. Technol.*, 1994: **28**, 53–62.

5. Grathwohl, P., Influence of organic matter from soils and sediments from various origins on the sorption of some chlorinated aliphatic hydrocarbons: implication on K_{oc} correlations: *Environ. Sci. Technol.*, 1990: **24**, 1687–1693.

6. Lyman, W. J., Adsorption Coefficient for Soils and Sediments, in *Handbook of Chemical Property Estimation Methods*, W. J. Lyman, W. F. Reehl, and D. H. Rosenblatt, Editors, 1990. Washington, DC: American Chemical Society.

7. Gawlik, B. M., et al., Alternatives for the Determination of the Soil Adsorption Coefficient, K_{oc}, of Non-ionic Organic Compounds—A Review. *Chemosphere*, 1997: **34**(12), 2525–2551.

8. Werth, C. J., and M. Reinhard, Effects of Temperature on Trichloroethylene Desorption from Silica Gel and Natural Sediments. 1. Isotherms. *Environ. Sci. Technol.*, 1997: **31**, 689–696.

9. Zachara, J. M., et al., Sorption of Binary Mixtures of Aromatic Nitrogen Heterocyclic Compounds on Subsurface Materials. *Environ. Sci. Technol.*, 1987: **21**, 397–402.

10. Schellenberg, K., C. Leuenberger, and R. P. Schwarzenbach, Sorption of Chlorinated Phenols by Natural Sediments and Aquifer Materials. *Environ. Sci. Technol.*, 1984: **18**, 652–657.

11. Lagas, P., Sorption of Chlorophenols in the Soil. *Chemosphere*, 1988: **17**, 205–216.

12. Lee, L. S., P. S. C. Rao, and M. L. Brusseau, Nonequilibrium Sorption and Transport of Neutral and Ionized Chlorophenols. *Environ. Sci. Technol.*, 1991: **25**, 722–729.

13. Meylan, W. M., P. H. Howard, and R. S. Boethling, Molecular Topology/Fragment Contribution Method for Predicting Soil Sorption Coefficients. *Environ. Sci. Technol.*, 1992: **26**, 1560–1567.

14. Wang, Z., D. S. Gamble, and C. H. Langford, Interaction of Atrazin with Laurentian Soil. *Environ. Sci. Technol.*, 1992: **26**, 560–565.

15. Lyman, W. J., W. F. Reehl, and D. H. Rosenblatt, *Handbook of Chemical Property Estimation Methods*, 1990. Washington, DC: American Chemical Society.

16. Sabljic, A., On the Prediction of Soil Sorption Coefficients of Organic Pollutants from Molecular Structure: Application of Molecular Topology Model. *Environ. Sci. Technol.*, 1987: **21**, 358–366.

17. Abdul, A. S., T. L. Gibson, and D. N. Rai, Statistical Correlations for Predicting the Partition Coefficient for Nonpolar Organic Contaminants Between Aquifer Organic Carbon and Water. *Hazard. Waste Hazard. Mater.*, 1987: **4**, 211–222.

18. Paya-Perez, A. B., M. Riaz, and B. R. Larsen, Soil Sorption of 20 PCB Congeners and Six Chlorobenzenes. *Ecotoxicol. Environ. Safety*, 1991: **21**, 1–17.

19. Vowles, P. D., and R. F. C. Mantoura, Sediment–Water Partition Coefficient and HPLC Retention Factors of Aromatic Hydrocarbons. *Chemosphere*, 1987: **16**, 109–116.

20. Hodson, J., and N. A. Williams, The Estimation of the Adsorption Coefficient (K_{oc}) for Soils by High Performance Liquid Chromatography. *Chemosphere*, 1988: **17**, 67–77.

21. He, Y., et al., Determination and Estimation of Physicochemical Properties for Phenyl-sulfonyl Acetates. *Chemosphere*, 1995: **30**, 117–125.

22. Bahnick, D. A., and W. J. Doucette, Use of Molecular Connectivity Indices to Estimate Soil Sorption Coefficients for Organic Chemicals. *Chemosphere*, 1988: **17**, 1703–1715.

23. Okouchi, S., and H. Saegusa, Prediction of Soil Sorption Coefficients of Hydrophobic Organic Pollutants by Adsorbability Index. *Bull. Chem. Soc. Jpn.*, 1989: **62**, 922–924.

24. Meylan, W. M., and P. H. Howard, *Soil/Sediment Adsorption Constant Program PCKOC*, 1992. Boca Raton, FL: Lewis Publishers.

25. Lohninger, H., Estimation of Soil Partition Coefficients of Pesticides from Their Chemical Structure. *Chemosphere*, 1994: **29**, 1611–1626.

APPENDIX A

SMILES NOTATION: BRIEF TUTORIAL

The Simplified Molecular Input Line Entry System (SMILES) is frequently used for computer-aided evaluation of molecular structures [1–3]. SMILES is widely accepted and computationally efficient because SMILES uses atomic symbols and a set of intuitive rules. Before presenting examples, the basic rules needed to enter molecular structures as SMILES notation are given.

Linear Notation A SMILES notation is a string consisting of alphanumeric and certain punctuation characters. The notation terminates at the first space encountered while reading sequentially from left to right.

Atoms Atoms present in typical organic compounds are called the *organic subset.* These atoms are represented by their atomic symbols:

$$B, C, N, O, P, S, F, Cl, Br, I$$

Hydrogen atoms are usually omitted unless they are required in certain cases. Aliphatic and nonaromatic carbon is indicated by the capital letter C. Atoms in aromatic rings are specified by lowercase letters. For example, an amino group is represented by the letter N, the nitrogen atom in a pyridine ring by n, a carbon in benzene or a pyridine ring as c. Atoms not included in the organic subset must be given in brackets (e.g., [Au]).

Bonds The symbols to specify bonds are as follows

single	–
double	=
triple	#
aromatic	:

Double and triple bonds must be indicated; single and aromatic bonds may be omitted.

Charges Attached hydrogens and charges are always specified in brackets. The number of attached hydrogens is shown by the symbol H followed by an optional digit.

[H+]	proton
[OH−]	hydroxyl anion
[OH3+]	hydronium cation
[Fe++]	iron(II) cation
[NH4+]	ammonium cation

Branches   Branches are represented by enclosure in parentheses. They can be nested or stacked.

Cyclic Structures Cyclic structures are represented by breaking one single or one aromatic bond in each ring. These bonds are numbered in any order, designating ring-opening bonds by a digit immediately following the atomic symbol at each breaking site. The remaining, noncyclic structure is denoted by using the rules above.

Examples

SMILES is based on the concept of hydrogen-suppressed molecular graphs (HSMG). The following example shows three representations of 1-butanol:

Formula	HO-CH2-CH2-CH2-CH3
HSMG	O-C-C-C-C
SMILES	OCCCC

The HSMG is derived simply by stripping the hydrogen atoms from the formula representation. The HSMG notation is already a valid SMILES notation. To obtain further compactness, the specification rules of SMILES allow the omission of single bonds. Although the final SMILES notation for 1-butanol is less than half as long as the formula notation, no information is lost since the complete molecular graph is reconstructed automatically by applying the valence rules. Exceptional cases where writing hydrogen atoms in the SMILES notation is required are described further below.

Now, encoding the branched isomers of 1-butanol with SMILES is straightforward:

2-Butanol	CC(O)CC	
iso-Butanol	OCC(C)C	
tert-Butanol	OC(C)(C)C	

Double bonds must be specified. Examples are given for ethene and its chlorinated derivatives:

Ethene	C=C	
Chloroethene (vinyl chloride)	ClC=C	
1,1-Dichloroethene	ClC(Cl)=C	
cis-1,2-Dichloroethene	ClC=CCl	

Note: The *cis* and *trans* isomers cannot be identified

Trichloroethene (TCE)	ClC(Cl)=CCl	
Perchloroethene (PCE)	ClC(Cl)=C(Cl)Cl	

The following examples show compounds with triple bonds:

Propyne	C#CC	$HC\equiv C-CH_3$
Propionitrile	N#CC	$N\equiv C-CH_3$

Examples of nonaromatic cyclic compounds are:

Cyclopropane	ClCCl	
Oxirane	OlCCl	
1,3-Dioxane	OlCOCCCl	
1,4-Dioxane	OlCCOCCl	
Morpholine	NlCCOCCl	
Cyclohexanol	OClCCCCCl	

Cyclohexanone   O=ClCCCCCl

2-Fluorocyclohexanone   O=ClC(F)CCCCl

3-Fluorocyclohexanone   O=ClCC(F)CCCl

4-Fluorocyclohexanone   O=ClCCC(F)CCl

Examples of aromatic compounds are:

Benzene   clccccc1

Pyridine   nlccccc1

s-Triazine   nlcncncl

Furane   olcccc1

Thiophene   slcccc1

Furfural   olccccclC=O

3-Cyanopyridine   nlcc(C#N)cccl

1,4-Bis(trifluoromethyl)benzene   FC(F)(F)clccc(C(F)(F)F)ccl

Unique SMILES Notation

In most cases there is more than one way to write the SMILES notation for a given compound. For example, pyridine can be entered in six different but correct ways:

(I) nlccccc1   (II) clncccc1   (III) clcnccc1
(IV) clccnccl   (V) clcccncl   (VI) clccccnl

To obtain a unique SMILES notation, computer programs such as the Toolkit include the CANGEN algorithm [1] which performs CANonicalization, resulting in unique enumeration of atoms, and then GENerates the unique SMILES notation for the canonical structure. In the case of pyridine, this is notation (III). Any molecular structure entered in the Toolkit is converted automatically into its unique representation.

REFERENCES

1. Weininger, D., SMILES, a Chemical Language and Information System. 1. Introduction to Methodology and Encoding Rules. *J. Chem. Inf. Comput. Sci.*, 1988: **28,** 31–36.
2. Weininger, D., A. Weininger, and J. L. Weininger, SMILES. 2. Algorithm for Generation of Unique SMILES Notation. *J. Chem. Inf. Comput. Sci.*, 1989: **29,** 97–101.
3. Weininger, D., SMILES. 3. DEPICT: Graphical Depiction of Chemical Structures. *J. Chem. Inf. Comput. Sci.*, 1990: **30,** 273–243.

APPENDIX B

DENSITY–TEMPERATURE FUNCTIONS

TABLE B.1   Hydrocarbons (Eq. 3.5.2, $a_3 = 0$ Unless Stated Otherwise)

Coefficients			Units			
a_0	a_1	a_2	ρ	T	T Range	Reference
		2,2-Dimethylpropane				
922.4	-1.26	0.451×10^{-3}	$\mathrm{kg\,m^{-3}}$	K	257–296 K	[1]
		trans-2,2,5,5-Tetramethyl-3-hexene				
1316	-2.981	3.21×10^{-3}	$\mathrm{kg\,m^{-3}}$	K	279–348 K	[1]
		1,6-Heptadiene				
950.5	-0.734	-0.297×10^{-3}	$\mathrm{kg\,m^{-3}}$	K	201–308 K	[1]
		1,7-Octadiene				
981.5	-0.985	0.108×10^{-3}	$\mathrm{kg\,m^{-3}}$	K	204–303 K	[1]
		1,3,5-Hexatriene				
1380	-3.532	4.537×10^{-3}	$\mathrm{kg\,m^{-3}}$	K	273–350 K	[1]
		Cycloheptane				
0.997387	-0.454838×10^{-3}	-0.619601×10^{-6}	$\mathrm{g\,cm^{-3}}$	°C	25–100°C	[2]
		Cyclodecane ($a_3 = -0.2018 \times 10^{-8}$)				
0.87340	-0.78424×10^{-3}	0.5005×10^{-6}	$\mathrm{g\,cm^{-3}}$	°C	25–192°C	[3]
		Bicyclo[2,2,2]octane				
1610	-2.693	1.795×10^{-3}	$\mathrm{kg\,m^{-3}}$	K	455–498 K	[1]

TABLE B.1 *(continued)*

Coefficients			Units			
a_0	a_1	a_2	ρ	T	T Range	Reference
		1,2,4,5-Tetramethylbenzene				
1070	-0.517	0.389×10^{-3}	$\mathrm{kg\,m^{-3}}$	K	358–446 K	[1]
		Hexamethylbenzene				
1344	-1.496	0.744×10^{-3}	$\mathrm{kg\,m^{-3}}$	K	454–512 K	[1]
		Anthracene				
1253	-0.355	-0.446×10^{-3}	$\mathrm{kg\,m^{-3}}$	K	490–558 K	[1]

TABLE B.2 **Halomethanes (Eq. 3.5.2, $a_3 = 0$ Unless Stated Otherwise)**

Coefficients			Units			
a_0	a_1	a_2	ρ	T	T Range	Reference
		Monofluoromethane				
1243	-1.747	-0.706×10^{-3}	$\mathrm{kg\,m^{-3}}$	K	139–219 K	[1]
		Difluoromethane				
1693	-1.801	-1.862×10^{-3}	$\mathrm{kg\,m^{-3}}$	K	159–222 K	[1]
		Trifluoromethane				
2062	-2.899	-2.007×10^{-3}	$\mathrm{kg\,m^{-3}}$	K	124–218 K	[1]
		Monobromomethane				
2.3232	-1.7037×10^{-3}	-0.01722×10^{-4}	$\mathrm{g\,cm^{-3}}$	K	178–323 K	[4]
		Dibromomethane				
3.1840	-2.0759×10^{-3}	-0.009142×10^{-4}	$\mathrm{g\,cm^{-3}}$	K	273–373 K	[4]
		Tribromomethane				
3.5603	-1.9668×10^{-3}	-0.01080×10^{-4}	$\mathrm{g\,cm^{-3}}$	K	283–395 K	[4]
		Tetrabromomethane ($a_3 = 0.5261 \times 10^{-7}$)				
-1.79705	34.5153×10^{-3}	0.78002×10^{-4}	$\mathrm{g\,cm^{-3}}$	K	373–463 K	[4]
		Monoiodomethane				
2.9124	-1.4918×10^{-3}	-0.02279×10^{-4}	$\mathrm{g\,cm^{-3}}$	K	273–313 K	[4]

TABLE B.2   *(continued)*

Coefficients			Units			
a_0	a_1	a_2	ρ	T	T Range	Reference

Diiodomethane

4.2314	-3.5538×10^{-3}	-0.01531×10^{-4}	$g\,cm^{-3}$	K	288–393 K	[4]

Fluorochloromethane ($a_3 = -0.136 \times 10^{-7}$)

2.0156	-4.0559×10^{-3}	0.08306×10^{-4}	$g\,cm^{-3}$	K	193–313 K	[4]

Difluorochloromethane ($a_3 = -0.096709$)

0.93746	0.22082	0.33821	$g\,cm^{-3}$	K	193–313 K	[4]

Trifluorochloromethane ($a_3 = 0.73578$)

0.93484	0.81088	0.906328	$g\,cm^{-3}$	K	193–313 K	[4]

Fluorodichloromethane ($a_3 = 0.99779$)

0.77124	1.1746	-1.4790	$g\,cm^{-3}$	K	193–313 K	[4]

Fluorotrichloromethane ($a_3 = 0.48140$)

0.92841	0.62942	-0.52335	$g\,cm^{-3}$	K	193–313 K	[4]

Trifluorobromomethane ($a_3 = 0.98925$)

1.0978	0.14426	-1.5083	$g\,cm^{-3}$	K	193–313 K	[4]

TABLE B.3   Halogenated Benzenes and Biphenyles (Eq. 3.5.2, $a_3 = 0$)

Coefficients			Units			
a_0	a_1	a_2	ρ	T	T Range	Reference

Bromobenzene

1.74269	-0.430199×10^{-3}	-1.42156×10^{-6}	$g\,cm^{-3}$	°C	25–100°C	[2]

2-Bromobiphenyl

1.82039	-1.81145×10^{-3}	-1.25885×10^{-6}	$g\,cm^{-3}$	K	253–333 K	[5]

2-Iodobiphenyl

2.07432	-1.94711×10^{-3}	-1.26361×10^{-6}	$g\,cm^{-3}$	K	253–333 K	[5]

TABLE B.4   Alcohols (Eq. 3.5.2, $a_3 = 0$ Unless Stated Otherwise)

Coefficients			Units			
a_0	a_1	a_2	ρ	T	T Range	Reference
Ethanol [ln $\rho(T)$ instead of $\rho(T)$, $a_3 = -0.747 \times 10^{-8}$]						
-0.215231	-1.05037×10^{-3}	-0.7837×10^{-6}	$\mathrm{g\,cm^{-3}}$	°C	-50 to 20°C	[6]
3-Methyl-1-butanol						
0.96907	-0.358821×10^{-3}	-0.610029×10^{-6}	$\mathrm{g\,cm^{-3}}$	K	298–343 K	[7]
1-Nonanol						
0.927999	$-0.0325735 \times 10^{-3}$	-1.05093×10^{-6}	$\mathrm{g\,cm^{-3}}$	°C	25–100°C	[2]
2-Methylcyclohexanol						
1.01049	0.139793×10^{-3}	-1.49594×10^{-6}	$\mathrm{g\,cm^{-3}}$	°C	25–100°C	[2]
2-Methyl-3-buten-2-ol						
1.11667	-0.0009893	0	$\mathrm{g\,cm^{-3}}$	K	293–353 K	[8]
Benzyl alcohol						
1.20061×10^{-3}	-0.319040×10^{-6}	-0.710885	$\mathrm{g\,cm^{-3}}$	K	298–343 K	[7]

TABLE B.5   Ethanolamines (Eq. 3.5.2, $a_3 = 0$)

Coefficients			Units			
a_0	a_1	a_2	ρ	T	T Range	Reference
Monoethanolamine						
1181.9	-0.38724	-6.1668×10^{-4}	$\mathrm{kg\,m^{-3}}$	K	294–432 K	[9]
Diethanolamine						
1212.0	-0.17861	-7.2922×10^{-4}	$\mathrm{kg\,m^{-3}}$	K	297–433 K	[9]
Triethanolamine						
1233.2	-0.20236	-5.8608×10^{-4}	$\mathrm{kg\,m^{-3}}$	K	295–435 K	[9]
N,N-Dimethylethanolamine						
1088.1	-0.50208	-6.1169×10^{-4}	$\mathrm{kg\,m^{-3}}$	K	299–398 K	[9]
N,N-Diethylethanolamine						
1071.4	-0.41388	-7.7194×10^{-4}	$\mathrm{kg\,m^{-3}}$	K	296–413 K	[9]

TABLE B.5 *(continued)*

	Coefficients			Units			
a_0	a_1	a_2		ρ	T	T Range	Reference
N-Methyldiethanolamine							
1207.0	-0.43265	-4.7744×10^{-4}		kg m^{-3}	K	296–471 K	[9]
N-Ethyldiethanolamine							
1209.4	-0.57829	-2.9971×10^{-4}		kg m^{-3}	K	294–472 K	[9]

TABLE B.6 Miscellaneous Compounds (Eq. 3.5.2, $a_3 = 0$)

	Coefficients			Units			
a_0	a_1	a_2		ρ	T	T Range	Reference
2-Bromo-2-chloro-1,1,1-trifluoroethane (halothane)							
2685.52	-2.7856	0		kg m^{-3}	K	291–314 K	[10]
2-Methyltetrahydrofurane							
1157	-1.124	-0.254×10^{-3}		kg m^{-3}	K	200–353 K	[1]
1,4-Dioxan							
1.32842	-0.893913×10^{-3}	-0.381851×10^{-6}		g cm^{-3}	K	298–343 K	[7]
Benzaldehyde							
1.38467	-1.34763×10^{-3}	-0.663290×10^{-6}		g cm^{-3}	°C	25–100°C	[2]
Ethyl Acetate							
1.15421	-0.513149×10^{-3}	-1.20242×10^{-6}		g cm^{-3}	K	298–343 K	[7]
2-Butoxyethanol							
0.91694	-0.8149×10^{-3}	-0.51×10^{-6}		g cm^{-3}	°C	20–60°C	[11]
Dimethylformamide							
0.9688	-0.94×10^{-3}	-0.73×10^{-6}		g cm^{-3}	°C	-10–40°C	[12]
Trichloroethene							
1498.14	-1.6777	0		kg m^{-3}	°C	18–42°C	[13]
Diphenyl-p-isopropylphenyl phosphate							
1.1786	-0.000794	0		g cm^{-3}	°C	29–241°C	[14]

REFERENCES

1. Shinsaka, K., N. Gee, and G. R. Freeman, Densities Against Temperature of 17 Organic Liquids and of Solid 2,2-Dimethylpropane. *J. Chem. Thermodyn.*, 1985: **17**, 1111–1119.
2. Nayar, S., and A. P. Kudchadker, Densities of Some Organic Substances. *J. Chem. Eng. Data*, 1973: **18**, 356–357.
3. McLure, I. A., and J.-M. Barbarin-Castillo, Density of Cyclodecane from to 192°C. *J. Chem. Eng. Data*, 1985: **30**, 253–254.
4. Kudchadker, A. P., et al., Densities of Liquid CH_4-aXa (X = Br,I) and CH_4-$(a+b+c+d)$ FaClbBrcId Halomethanes. *J. Phys. Chem. Ref. Data*, 1978: **7**, 425–439.
5. Amey, L., and A. P. Nelson, Densities and Viscosities of 2-Bromobiphenyl and 2-Iodobiphenyl. *J. Chem. Eng. Data*, 1982: **27**, 253–254.
6. Schroeder, M. R., B. E. Poling, and D. B. Manley, Ethanol Densities Between −50 and 20°C. *J. Chem. Eng. Data*, 1982: **27**, 256–258.
7. Abraham, T., V. Bery, and A. Kudchadker, Densities of Some Organic Substances. *J. Chem. Eng. Data,* 1971: **16**, 355–356.
8. Gubareva, A. I., P. A. Gerasimov, and V. V. Beregovykh, Physicochemical Properties of Dimethylvinylcarbinol (2-Methyl-3-buten-2-ol). *Zh. Prikl. Khim.*, 1984: **57EE**, 2383–2385.
9. DiGullio, R. M., et al., Densities and Viscosities of the Ethanolamines. *J. Chem. Eng. Data*, 1992: **37**, 239–242.
10. Francesconi, R., F. Comelli, and Giacomini, D., Excess Enthalpy for the Binary System 1,3-Dioxolane + Halothane. *J. Chem. Eng. Data*, 1990: **35**, 190–191.
11. McLure, I. A., F. Guzman-Figueroa, and I. L. Pegg, Density of 2-Butoxyethanol from 20 to 60°C. *J. Chem. Eng. Data*, 1982: **27**, 398–399.
12. Ennan, A. A., et al., Density and Viscosity in the System Dimethylformamide–Water and Water–Urotropin in the Temperature Range from − 15 to + 40°C. *Zh. Prikl. Khim.*, 1972: **45EE**, 627–630.
13. Francesconi, R., and F. Comelli, Heat of Mixing of 1,3-Dioxolane + Trichloroethylene. *J. Chem. Eng. Data*, 1990: **35**, 291–293.
14. Tarasov, I. V., Y. V. Golubkov, and R. I. Luchkina, Density, Viscosity, and Surface Tension of Diphenyl-p-Isopropylphenyl Phosphate. Zh. Prikl. Khim., 1982: **56EE**, 2578–2580.

APPENDIX C

VISCOSITY–TEMPERATURE FUNCTIONS

TABLE C.1   Hydrocarbons (Eq. 6.4.3)

a_0	a_1	a_2	Viscosity	T	T Range	Reference
	Coefficients		Units			
n-Hexane ($a_3 = 0$)						
-5.33644	1585.81	-0.1070	η (mPa·s)	K	288–326°C	[1]
n-Heptane ($a_3 = 0$)						
-5.19395	1599.43	-0.09903	η (mPa·s)	K	293–333°C	[1]
n-Octane ($a_3 = 117.3 \times 10^{-6}$)						
-8.19695	4751.80	-1.14146	η (mPa·s)	K	293–346°C	[1]
n-Decane ($a_3 = 141.3 \times 10^{-6}$)						
-8.01441	4969.45	-1.25766	η (mPa·s)	K	293–423°C	[1]
n-Dodecane ($a_3 = 159.0 \times 10^{-6}$)						
-7.74170	5019.23	-1.31439	η (mPa·s)	K	293–425°C	[1]
n-Tetradecane ($a_3 = 159.5 \times 10^{-6}$)						
-7.37053	4808.83	-1.24841	η (mPa·s)	K	293–423°C	[1]
Cyclohexane ($a_3 = 0$)						
-4.15991	911.472	-0.08807	η (mPa·s)	K	288–333°C	[1]
Benzene ($a_3 = 0$)						
-3.37475	442.965	-0.1228	η (mPa·s)	K	288–333°C	[1]

TABLE C.2   Alkanols (Eq. 6.4.2)

Coefficients			Units			
b_1	b_2	b_3	Viscosity	T	T Range	Reference
			Methanol			
-4.7928	1247.6	0	η (cP)	K	288–328 K	[2]
			Ethanol			
-5.5406	1667.3	0	η (cP)	K	288–328 K	[2]
			1-Propanol			
-6.1721	2022.1	0	η (cP)	K	288–328 K	[2]

TABLE C.3   Butanediols (Eq. 6.4.2)

Coefficients			Units			
b_1	b_2	b_3	Viscosity	T	T Range	Reference
			1,2-Butanediol			
1519.054	-3.435384	0.008163260	η (mPa·s)	K	301–454 K	[3]
			1,3-Butanediol			
1323.893	-1.787522	0.003560611	η (mPa·s)	K	302–459 K	[3]
			1,4-Butanediol			
1396.844	-2.432001	0.005549931	η (mPa·s)	K	303–461 K	[3]
			2,3-Butanediol			
1375.06	2.102106	0.003962518	η (mPa·s)	K	303–461 K	[3]

TABLE C.4   Ethanolamines (Eq. 6.4.2)

Coefficients			Units			
b_1	b_2	b_3	Viscosity	T	T Range	Reference
			Monoethanolamine			
-3.9356	1010.8	151.17	η (mPa·s)	K	303–424 K	[4]
			Diethanolamine			
-5.2380	1672.9	153.82	η (mPa·s)	K	292–424 K	[4]

TABLE C.4 *(continued)*

Coefficients			Units			
b_1	b_2	b_3	Viscosity	T	T Range	Reference
			Triethanolamine			
− 3.5957	1230.3	175.35	η (mPa·s)	K	293–424 K	[4]
			N,N-Dimethylethanolamine			
− 5.2335	1453.6	71.773	η (mPa·s)	K	294–394 K	[4]
			N,N-Diethylethanolamine			
− 4.2337	884.19	141.15	η (mPa·s)	K	295–414 K	[4]
			N-Methyldiethanolamine			
− 4.3039	1266.2	151.40	η (mPa·s)	K	293–424 K	[4]
			N-Ethyldiethanolamine			
− 3.9927	1090.8	− 164.21	η (mPa·s)	K	292–424 K	[4]

REFERENCES

1. Knapstad, B., P. A. Skjølsvik, and H. A. Øye, Viscosity of Pure Hydrocarbons. *J. Chem. Eng. Data,* 1989: **34**, 37–43.
2. Crabtree, A. M., and J. F. O'Brien, Excess Viscosities of Binary Mixtures of Chloroform and Alcohols. *J. Chem. Eng. Data*, 1991: **36**, 140–142.
3. Sun, T., R. M. DiGullo, and A. S. Teja, Densities and Viscosities of Four Butanediols Between 293 and 463. *K. J. Chem. Eng. Data*, 1992: **37**, 246–248.
4. DiGullio, R. M., et al., Densities and Viscosities of the Ethanolamines. *J. Chem. Eng. Data*, 1992: **37**, 239–242.

APPENDIX D

AWPC–TEMPERATURE FUNCTIONS

TABLE D.1   AWPC–Temperature Functions (Eq. 12.5.1) for Alkanes and Cycloalkanes

Y	A	B	Range (°C)	Reference
n-Hexane				
$\ln[H_c(\text{atm}\cdot\text{m}^3\,\text{mol}^{-1})]$	25.25	7530	10–30	[1]
n-Heptane				
$\ln[H_c(\text{kPa}\cdot\text{m}^3\,\text{mol}^{-1})]$	17 ± 2.22	3730 ± 686	25–45	[2]
n-Octane				
$\ln[H_c(\text{kPa}\cdot\text{m}^3\,\text{mol}^{-1})]$	30 ± 5.25	8014 ± 1617	25–45	[2]
2-Methylpentane				
$\ln[H_c(\text{atm}\cdot\text{m}^3\,\text{mol}^{-1})]^{\,a}$	2.959	957.2	10–30	[1]
2-Methylhexane				
$\ln[H_c(\text{kPa}\cdot\text{m}^3\,\text{mol}^{-1})]$	-8 ± 3.53	-3608 ± 1088	25–45	[2]
Cyclopentane				
$\ln[H_c(\text{kPa}\cdot\text{m}^3\,\text{mol}^{-1})]$	14 ± 2.03	3351 ± 633	25–45	[2]
Cyclohexane				
$\ln[H_c(\text{atm}\cdot\text{m}^3\,\text{mol}^{-1})]$	9.141	3238	10–30	[1]
Methylcyclohexane				
$\ln[H_c(\text{kPa}\cdot\text{m}^3\,\text{mol}^{-1})]$	34 ± 3.39	9406 ± 1046	25–45	[2]
Decalin (cis or trans isomer not specified)				
$\ln[H_c(\text{atm}\cdot\text{m}^3\text{mol}^{-1})]$	11.85	4125	10–30	[1]

$^a r^2 = 0.497.$

TABLE D.2   AWPC–Temperature Functions (Eq. 12.5.1) for Aromatic Hydrocarbons

Y	A	B	Range (°C)	Reference
		Benzene		
$\ln[H_c(\text{atm}\cdot\text{m}^3\,\text{mol}^{-1})]$	5.534	3194	10–30	[1]
$\ln[H_y(\text{-})]$	19.02	3964	0–30	[3]
		Toluene		
$\ln[H_c(\text{atm}\cdot\text{m}^3\,\text{mol}^{-1})]$	5.133	3024	10–30	[1]
$\ln]H_y(\text{-})]$	18.46	3751	0–30	[3]
		Ethylbenzene		
$\ln[H_c(\text{atm}\cdot\text{m}^3\,\text{mol}^{-1})]$	11.92	4994	10–30	[1]
		1,2-Dimethylbenzene		
$\ln[H_c(\text{atm}\cdot\text{m}^3\,\text{mol}^{-1})]$	5.541	3220	10–30	[1]
		1,3-Dimethylbenzene		
$\ln[H_c(\text{atm}\cdot\text{m}^3\,\text{mol}^{-1})]$	6.280	3337	10–30	[1]
		1,4-Dimethylbenzene		
$\ln[H_c(\text{kPa}\cdot\text{m}^3\,\text{mol}^{-1})]$	10 ± 0.56	3072 ± 173	25–45	[2]
$\ln[H_c(\text{atm}\cdot\text{m}^3\,\text{mol}^{-1})]$	6.931	3520	10–30	[1]
		1,2,4-Trimethylbenzene		
$\ln[H_c(\text{kPa}\cdot\text{m}^3\,\text{mol}^{-1})]$	14 ± 2.24	4298 ± 686	25–45	[2]
		1,3,5-Trimethylbenzene		
$\ln[H_c(\text{atm}\cdot\text{m}^3\,\text{mol}^{-1})]$	7.241	3628	10–30	[1]
		n-Propybenzene		
$\ln[H_c(\text{atm}\cdot\text{m}^3\,\text{mol}^{-1})]$	7.835	3681	10–30	[1]
		Methyl ethylbenzene (o, m, or p isomer not specificied)		
$\ln[H_c(\text{atm}\cdot\text{m}^3\,\text{mol}^{-1})]$	5.557	3179	10–30	[1]
		Cumene		
$\ln[H_c(\text{kPa}\cdot\text{m}^3\,\text{mol}^{-1})]$	11 ± 1.84	3269 ± 564	25–45	[2]
		Tetralin		
$\ln[H_c(\text{atm}\cdot\text{m}^3\,\text{mol}^{-1})]$	11.83	5392	10–30	[1]
		2-Methylnaphthalene		
$\ln[H_c(\text{kPa}\cdot\text{m}^3\,\text{mol}^{-1})]$	7 ± 0.14	1234 ± 44	25–45	[2]

TABLE D.3   AWPC–Temperature Functions (Eq. 12.5.1) for Halogenated Methanes

Y	A	B	Range (°C)	Reference
		Chloromethane		
$\ln[H_c(\text{atm}\cdot\text{m}^3\,\text{mol}^{-1})]$	9.358	4215	10–35	[4]
		Dichloromethane		
$\ln[H_c(\text{atm}\cdot\text{m}^3\,\text{mol}^{-1})]$	8.483	4268	10–30	[1]
$\ln[H_c(\text{atm}\cdot\text{m}^3\,\text{mol}^{-1})]$	6.653	3817	10–35	[4]
$\ln[H_y(\text{-})]$	17.42	3645	0–30	[3]
		Trichloromethane		
$\ln[H_c(\text{atm}\cdot\text{m}^3\,\text{mol}^{-1})]$	11.41	5030	10–30	[1]
$\ln[H_c(\text{atm}\cdot\text{m}^3\,\text{mol}^{-1})]$	9.843	4612	10–35	[4]
$\ln[H_c(\text{atm}\cdot\text{m}^3\,\text{mol}^{-1})]$	11.9	5200	10–30	[5]
$\log[K_{aw}(\text{-})]$	4.990	1729	10–30	[6]
$\ln[H_y(\text{-})]$	18.97	4046	0–30	[3]
		Tetrachloromethane		
$\ln[H_c(\text{kPa}\cdot\text{m}^3\,\text{mol}^{-1})]$	13 ± 0.74	3553 ± 230	25–45	[2]
$\ln[H_c(\text{atm}\cdot\text{m}^3\,\text{mol}^{-1})]$	9.739	3951	10–30	[1]
$\ln[H_c(\text{atm}\cdot\text{m}^3\,\text{mol}^{-1})]$	11.29	4411	10–35	[4]
$\log[K_{aw}(\text{-})]$	5.853	1718	10–30	[6]
$\ln[H_y(\text{-})]$	22.22	4438	0–30	[3]
		Dichlorodifluoromethane		
$\log[K_{aw}(\text{-})]$	5.811	1399	10–30	[6]
		Trichlorofluoromethane		
$\ln[H_c(\text{atm}\cdot\text{m}^3\,\text{mol}^{-1})]$	9.480	3513	10–30	[1]
		Tribromomethane		
$\ln[H_c(\text{atm}\cdot\text{m}^3\,\text{mol}^{-1})]$	11.6	5670	10–30	[5]
$\log[K_{aw}(\text{-})]$	4.729	1905	0–30	[6]
		Bromodichloromethane		
$\ln[H_c(\text{atm}\cdot\text{m}^3\,\text{mol}^{-1})]$	11.3	5210	10–30	[5]
		Dibromochloromethane		
$\ln[H_c(\text{atm}\cdot\text{m}^3\,\text{mol}^{-1})]$	14.62	6373	10–30	[1]
$\ln[H_c(\text{atm}\cdot\text{m}^3\,\text{mol}^{-1})]$	10.7	5210	10–30	[5]

TABLE D.4   AWPC–Temperature Functions (Eq. 12.5.1) for Halogenated Ethanes

Y	A	B	Range (°C)	Reference
Chloroethane				
$\ln[H_c(\text{atm} \cdot \text{m}^3 \, \text{mol}^{-1})]$	4.265	2580	10–30	[1]
$\ln[H_c(\text{atm} \cdot \text{m}^3 \, \text{mol}^{-1})]$	5.974	3120	10–35	[4]
1,1-Dichloroethane				
$\ln[H_c(\text{atm} \cdot \text{m}^3 \, \text{mol}^{-1})]$	5.484	3137	10–30	[1]
$\ln[H_c(\text{atm} \cdot \text{m}^3 \, \text{mol}^{-1})]$	8.637	4128	10–35	[4]
1,2-Dichloroethane				
$\ln[H_c(\text{atm} \cdot \text{m}^3 \, \text{mol}^{-1})]^a$	− 1.371	1522	10–30	[1]
$\ln[H_y(\text{-})]$	16.05	3539	0–30	[3]
1,1,1-Trichloroethane				
$\ln[H_c(\text{kPa} \cdot \text{m}^3 \, \text{mol}^{-1})]$	11 ± 0.30	3120 ± 93	25–45	[2]
$\ln[H_c(\text{atm} \cdot \text{m}^3 \, \text{mol}^{-1})]$	7.351	3399	10–30	[1]
$\ln[H_c(\text{atm} \cdot \text{m}^3 \, \text{mol}^{-1})]$	9.777	4133	10–35	[4]
$\log[K_{aw}(\text{-})]$	5.327	1636	10–30	[6]
$\ln[H_y(\text{-})]$	21.68	4375	0–30	[3]
1,1,2-Trichloroethane				
$\ln[H_c(\text{kPa} \cdot \text{m}^3 \, \text{mol}^{-1})]$	17 ± 3.36	5901 ± 1158	25–45	[2]
$\ln[H_c(\text{atm} \cdot \text{m}^3 \, \text{mol}^{-1})]$	9.320	4843	10–30	[1]
$\ln[H_y(\text{-})]$	16.20	3690	0–30	[3]
1,1,2,2-Tetrachloroethane				
$\ln[H_y(\text{-})]$	14.91	3547	0–30	[3]
Hexachloroethane				
$\ln[H_c(\text{atm} \cdot \text{m}^3 \, \text{mol}^{-1})]$	3.744	2550	10–30	[1]
$\ln[H_{aw}(\text{-})]$	6.982	2320	10–30	[6]
1,1,2-Trichlorotrifluoroethane				
$\ln[H_c(\text{atm} \cdot \text{m}^3 \, \text{mol}^{-1})]$	9.649	3243	10–30	[1]
1,2-Dibromoethane				
$\ln[H_c(\text{atm} \cdot \text{m}^3 \, \text{mol}^{-1})]$	5.703	3876	10–30	[1]

$^a r^2 = 0.878.$ $^b r^2 = 0.768.$

TABLE D.5 AWPC–Temperature Functions (Eq. 12.5.1) for Chlorinated Propanes, Butanes, Pentanes, and Hexanes

Y	A	B	Range (°C)	Reference
		1,2-Dichloropropane		
$\ln[H_c(\text{atm}\cdot\text{m}^3\,\text{mol}^{-1})]$	9.843	4708	10–30	[1]
$\ln[H_y(\text{-})]$	19.60	4333	0–30	[3]
		1,3-Dichloropropane		
$\ln[H_y(\text{-})]$	17.13	3917	0–30	[3]
		1,2,3-Trichloropropane		
$\ln[H_y(\text{-})]$	14.61	3477	0–30	[3]
		1-Chlorobutane		
$\ln[H_y(\text{-})]$	18.51	3482	0–30	[3]
		2-Chlorobutane		
$\ln[H_y(\text{-})]$	22.29	4499	0–30	[3]
		1,4-Dichlorobutane		
$\ln[H_y(\text{-})]$	13.79	3128	0–30	[3]
		1-Chloropentane		
$\ln[H_y(\text{-})]$	23.04	4727	0–30	[3]
		1,5-Dichloropentane		
$\ln[H_y(\text{-})]$	8.79	1597	0–30	[3]
		1-Chlorohexane		
$\ln[H_y(\text{-})]$	22.16	4459	0–30	[3]

TABLE D.6 AWPC–Temperature Functions (Eq. 12.5.1) for Chlorinated Ethenes

Y	A	B	Range (°C)	Reference
		Chloroethene		
$\ln[H_c(\text{atm}\cdot\text{m}^3\,\text{mol}^{-1})]$	6.138	2931	10–30	[1]
$\ln[H_c(\text{atm}\cdot\text{m}^3\,\text{mol}^{-1})]$	7.385	3286	10–35	[4]
		1,1-Dichloroethene		
$\ln[H_c(\text{atm}\cdot\text{m}^3\,\text{mol}^{-1})]$	6.123	2907	10–30	[1]
$\ln[H_c(\text{atm}\cdot\text{m}^3\,\text{mol}^{-1})]$	8.845	3729	10–35	[4]
$\ln[H_y(\text{-})]$	23.12	4618	0–30	[3]

TABLE D.6   *(continued)*

Y	A	B	Range (°C)	Reference
	cis-1,2-Dichloroethene			
$\ln[H_c(\text{atm} \cdot \text{m}^3 \text{ mol}^{-1})]$	5.164	3143	10–30	[1]
$\ln[H_c(\text{atm} \cdot \text{m}^3 \text{ mol}^{-1})]$	8.479	4192	10–35	[4]
	trans-1,2-Dichloroethene			
$\ln[H_c(\text{kPa} \cdot \text{m}^3 \text{ mol}^{-1})]$	11 ± 1.96	3396 ± 602	25–45	[2]
$\ln[H_c(\text{atm} \cdot \text{m}^3 \text{ mol}^{-1})]$	5.333	2964	10–30	[1]
$\ln[H_c(\text{atm} \cdot \text{m}^3 \text{ mol}^{-1})]$	9.341	4182	10–35	[4]
	Trichloroethene			
$\ln[H_c(\text{atm} \cdot \text{m}^3 \text{ mol}^{-1})]$	7.845	3702	10–30	[1]
$\ln[H_c(\text{atm} \cdot \text{m}^3 \text{ mol}^{-1})]$	11.37	4780	10–35	[4]
$\ln[H_{\text{aw}}(\text{-})]$	6.026	1909	10–30	[6]
$\ln[H_y(\text{-})]$	21.89	4647	0–30	[3]
	Tetrachloroethene			
$\ln[H_c(\text{atm} \cdot \text{m}^3 \text{ mol}^{-1})]$	10.65	4368	10–30	[1]
$\ln[H_c(\text{atm} \cdot \text{m}^3 \text{ mol}^{-1})]$	12.45	4918	10–35	[4]
$\ln[H_{\text{aw}}(\text{-})]$	5.920	1802	10–30	[6]
$\ln[H_y(\text{-})]$	22.68	4735	0–30	[3]

TABLE D.7   AWPC–Temperature Functions (Eq. 12.5.1) for Halogenated Benzenes and Alkylbenzenes

Y	A	B	Range (°C)	Reference
	Chlorobenzene			
$\ln[H_c(\text{atm} \cdot \text{m}^3 \text{ mol}^{-1})]$	3.469	2689	10–30	[1]
$\ln[H_y(\text{-})]$	16.83	3466	0–30	[3]
	o-Chlorotoluene			
$\ln[H_y(\text{-})]$	17.18	3545	0–30	[3]
	1,2-Dichlorobenzene			
$\ln[H_c(\text{atm} \cdot \text{m}^3 \text{ mol}^{-1})]^a$	-1.518	1422	10–30	[1]
	1,3-Dichlorobenzene			
$\ln[H_c(\text{atm} \cdot \text{m}^3 \text{ mol}^{-1})]^b$	2.882	2564	10–30	[1]
	1,4-Dichlorobenzene			
$\ln[H_c(\text{atm} \cdot \text{m}^3 \text{ mol}^{-1})]$	3.373	2720	10–30	[1]

TABLE D.7   *(continued)*

Y	A	B	Range (°C)	Reference
1,2,4-Trichlorobenzene				
$\ln[H_c(\text{atm} \cdot \text{m}^3 \text{ mol}^{-1})]^c$	7.361	4028	10–30	[1]
Bromobenzene				
$\ln[H_c(\text{kPa} \cdot \text{m}^3 \text{ mol}^{-1})]$	16 ± 1.11	5341 ± 346	25–45	[2]

[a] $r^2 = 0.464$.
[b] $r^2 = 0.850$.
[c] $r^2 = 0.819$.

TABLE D.8   **AWPC-Temperature Functions (Eq. 12.5.1) for Miscellaneous Compounds**

Y	A	B	Range (°C)	Reference
2-Butanone				
$\ln[H_c(\text{atm} \cdot \text{m}^3 \text{ mol}^{-1})]^a$	-26.32	-5214	10–30	[1]
2,4-Dimethylphenol				
$\ln[H_c(\text{atm} \cdot \text{m}^3 \text{ mol}^{-1})]^b$	-16.34	-3307	10–30	[1]

[a] $r^2 = 0.797$.
[b] $r^2 = 0.555$.

REFERENCES

1. Asworth, R. A., et al., Air–Water Partitioning Coefficients of Organics in Dilute Aqueous Solutions. *J. Hazard. Mater.,* 1988: **18**, 25–36.

2. Hansen, K. G., Z. Zhou, Yaws, C. L., and T. M. Aminabhavi, Determination of Henry's Law Constants orf Organic in Dilute Aqueous Solutions. *J. Chem. Eng. Data*, 1993: **38**, 546–550.

3. Leighton, D. T. J., and J. M. Calo, Distribution Coefficients of Chlorinated Hydrocarbons in Dilute Air–Water Systems for Groundwater Contamination Applications. *J. Chem. Eng. Data*, 1981: **26**, 382–385.

4. Gossett, J. M. Measurement of Henry's Law Constant for C1 and C2 Chlorinated Hydrocarbons. *Environ. Sci. Technol.*, 1987: **21**, 202–208.

5. Nicholson, B. C., B. P. Maguire, and D. B. Bursill, Henry's Law Constants for the Trihalomethanes: Effect of Water Composition and Temperature. *Environ. Sci. Technol.*, 1984: **18**, 518–521.

6. Munz, C., and P. V. Roberts, Air–Water Phase Equilibria of Volatile Organic Solutes. *J. Am. Water Works Assoc.*, 1987: May, 62–69.

APPENDIX E

CONTRIBUTION VALUES TO LOG *S* OF GROUP PARAMETERS IN MODELS OF KLOPMAN, WANG, AND BALTHASAR [1]

TABLE E.1   Model Ia

	Parameter[a]	Contribution	Remarks
	Constant C_o	3.5650	
1	$-CH_3$	-0.3361	
2	$-CH_2-$	-0.5729	
3	$-CH-(-)$	-0.6057	
4	$-C-(-)(-)$	-0.7853	
5	$=CH2$	-0.6870	
6	$=CH-$	-0.3230	
7	$=C-(-)$	-0.3345	
8	$=C=$	Undetermined	
9	$-C\equiv CH$	-0.6013	
10	$\equiv C-$	Undetermined	Not$-C\equiv N$, not in $-C\equiv CH$
11	$-C^*H_2-$	-0.4568	
12	$-C^*H-(-)$	-0.4072	
13	$-C^*-(-)(-)$	-0.3122	
14	$=C^*H-$	-0.3690	
15	$=C^*-(-)$	-0.4944	
16	$=C^*=$	Undetermined	
17	$\equiv C^*-$	Undetermined	
18	$-F$	-0.4472	Connecting to an sp^3 carbon
19	$-F$	-0.1773	Connecting to another atom
20	$-Cl$	-0.4293	Connecting to sp^3 carbon
21	$-Cl$	-0.6318	Connecting to another atom
22	$-Br$	-0.6321	Connecting to sp^3 carbon
23	$-Br$	-0.9643	Connecting to another atom
24	$-I$	-1.2391	Connecting to sp^3 carbon

TABLE E.1   *(continued)*

	Parameter[a]	Contribution	Remarks
25	$-I$	-1.2597	Connecting to another atom
26	$-OH$	1.4642	Primary alcohol
27	$-OH$	1.5629	Secondary alcohol
28	$-OH$	1.0885	Tertiary alcohol
29	$-OH$	1.1919	Connecting to a non-ap^3 carbon not in COOH
30	$-OH$	Undetermined	Connecting to a nitrogen
31	$-OH$	Undetermined	Connecting to a oxygen
32	$-OH$	Undetermined	Connecting to a phosphorus
33	$-OH$	Undetermined	Connecting to a sulfur
34	$-O^*-$	-0.2991	
35	$-O-$	0.8515	
36	$-CHO-$	0.4476	Aldehyde group
37	$-COOH$	0.2653	Conjugated acid
38	$-COOH$	1.1695	Nonconjugated acid
39	$-COO-$	0.8724	Ester
40	$-CONH_2$	Undetermined	
41	$-CONH-$	0.1931	
42	$-CON-(-)$	Undetermined	
43	$-CON=$	Undetermined	
44	$-CO-$	1.3049	
45	$-C^*O-$	1.5413	
46	$-NO-$	Undetermined	Not in NO_2
47	$-PO-$	Undetermined	
48	$-SO-$	0.5826	
49	$-NH_2$	0.6935	
50	$-NH-$	0.9549	
51	$-N^*H-$	Undetermined	
52	$-N-(-)$	Undetermined	
53	$-N^*-(-)$	Undetermined	
54	$-C\equiv N$	0.6262	
55	$HN=$	Undetermined	
56	$-N=$	Undetermined	
57	$-N^*=$	-0.3722	
58	$-NO_2$	-0.2647	
59	$-SH$	-0.5118	
60	$-S-$	Undetermined	
61	$-S^*-$	Undetermined	
62	$S=P-$	-2.4086	
63	$S=$	-1.3197	Not in $S=P$
64	P	Undetermined	sp^3 phosphorus
65	P	Undetermined	Non-sp^3 phosphorus, not in $S=P$
66	Alkanes	-1.5387	
67	Other hydrocarbons	-0.2598	Any hydrocarbon except for alkanes

[a]*indicates that the atom is in a ring system. The open valences in parmeters 1–65 are not filled by hydrogens.

Source: Reprinted with permission from Ref. 1. Copyright 1992. American Chemical Society.

TABLE E.2 Model IIa

	Parameter[a]	Contribution	Remarks
	Constant C_0	3.7253	
1	CH_3	−0.4169	sp^3
2	CH_2	−0.5199	sp^3
3	CH	−0.3057	sp^3
4	C	−0.1616	sp^3
5	$=CH_2$	−0.7788	sp^2
6	CH=	−0.3843	sp^2
7	C=	−0.5085	sp^2
8	C or CH	−0.4711	sp^2
9	NH_2	0.6184	sp^3
10	NH	0.7796	sp^3
11	N^*H	0.7974	sp^3
12	N	1.0734	sp^3
13	N^*	0.3906	sp^3
14	N=	−0.8015	sp^2
15	$N^*=$	−0.3677	sp^2
16	N	1.0026	sp
17	NO_2	−2.2003	
18	OH	1.0910	sp^3
19	O	0.4452	sp^3
20	O=	0.9545	sp^3
21	$S(sp^3)$	−0.6161	sp^3
22	$S(sp^2)$	−0.3648	sp^2
23	S ()	−1.9783	Another sulfur
24	P	−0.9139	Any phosphorus
25	F	−0.5862	
26	Cl	−0.6292	
27	Br	−0.9190	
28	I	−1.4676	
29	COO	−0.4537	An ester group
30	COOH	−1.2440	An acid group
31	$CONH_n (n=0,1,2)$	−0.5531	
32	Alkanes	−1.8549	
33	Other hydrocarbons	−0.2168	Except for alkanes

[a]*indicates that the atom is in a ring system.
Source: Reprinted with permission from Ref. 1. Copyright 1992. American Chemical Society.

REFERENCE

1. Klopman, G., S. Wang, and D. M. Balthasar, Estimation of Aqueous Solubility of Organic Molecules by the Group Contribution Approach: Application to the Study of Biodegradation *J. Chem. Inf. Comput. Sci.*, 1992: **32**, 474–482.

K_{ow} ATOM CONTRIBUTIONS OF BROTO, MOREAU, AND VANDYCKE

Key: H atoms: not shown; aromatic bonds: ⁗⁗; Nitrogroups: O–N=O; Sulfones, sulfoxides, and sulfonic derivatives: O–S–R

CI No.	Atom with Environment	Mean Value[a,b]	Std. Dev.[b]
1	$-\overset{\vert}{\underset{\vert}{C}}-$	0.198	0.066
2	$-C-$	0.456	0.009
3	$C-$	0.631	0.016
4	$\overset{\diagdown}{\underset{\diagup}{C}}-$	0.029	0.036
5	$\overset{\diagdown}{\underset{\diagup}{C}}{}^{⁗}$	0.095	0.023
6	$\overset{⁗}{C}{}^{⁗}$	0.311	0.006
7	$-C\equiv$	0.147	0.153
8	$C\equiv$	0.326	0.230
9	$\overset{\diagdown}{C}=$	0.099	0.032
10	$C=$	0.743	0.070
11	$\overset{⁗}{\underset{\shortmid}{C}}{}^{⁗}$	0.249	0.034
12	$=C=$	-0.036	0.452
13	$\overset{\diagdown}{\underset{\diagup}{C}}=$	-0.548	0.060

CI No.	Atom with Environment	Mean Value[a,b]	Std. Dev.[b]
14	$O{=}C\diagdown^{\diagup}$	− 0.681	0.062
15	$O{-}C\diagdown^{\sslash}$	0.532	0.058
16	$O{=}C\diagdown$	− 0.676	0.105
17	$O{-}C\diagdown$	− 0.990	0.065
18	$O{-}C\diagdown^{\diagup}$	− 0.936	0.122
19	$O{-}C\diagdown^{\diagup}{-}$	− 1.494	0.195
20	$O{-}C''''$	− 0.173	0.041
21	$N''''C\diagdown$	− 0.695	0.163
22	$N{-}C\diagdown$	− 1.086	0.091
23	$N{-}C\diagdown^{\diagup}$	− 1.372	0.136
24	$N{-}C\diagdown^{\diagup}{-}$	− 1.691	0.435
25	$N{-}C''''$	− 0.597	0.053
26	$N{-}Z$	− 0.285	0.113
27	$N{-}C\diagdown^{\sslash}$	− 0.110	0.053
28	$N{=}Z$	—	—
29	$Cl{-}C\diagdown$	0.309	0.088
30	$Cl{-}C''''$	0.835	0.031
31	$Br{-}C''''$	1.104	0.064
32	$F{-}C''''$	0.352	0.053
33	$O{=}Z$	0.339	0.272
34	$O{-}Z$	− 0.717	0.208
35	$F{-}C\diagdown^{\diagup}{-}$	0.333	0.033
36	$I{-}C''''$	1.511	0.099
37	$Cl{-}C\diagdown^{\diagup}$	0.605	0.046

CI No.	Atom with Environment	Mean Value[a,b]	Std. Dev.[b]
38	Cl—C⫽	0.841	0.111
39	Cl—C⟨	0.382	0.043
40	Br—C⫽	1.083	0.120
41	I—C⫽	1.208	0.147
42	F—C⟨	0.035	0.180
43	F—C⫽	0.432	0.166
44	Br—C⟍	0.620	0.193
45	F—C⟍	− 0.134	0.258
46	I—C⟍	0.893	0.305
47	Cl—C⫽	1.158	0.363
48	Br—C⟨	0.638	0.125
49	I—C⟨	0.802	0.333
50	I—Z	− 3.356	0.459
51	N=C⟨	− 0.836	0.281
52	N=Z	1.521	0.320
53	S=C⟨	− 0.174	0.113
54	S=C=	—	—
55	Br—C⟨	—	—
56	S—C⟍	0.278	0.431
57	S—C⫲	0.779	0.275
58	F—Z	1.739	0.261
59	S—C⫽	1.028	0.433
60	C—O—C	− 0.994	0.088
61	=C—O—C	− 0.515	0.311

CI No.	Atom with Environment	Mean Value[a,b]	Std. Dev.[b]
62	=C(/)–O–C\	0.201	0.047
63	=C(/)–O–C	0.371	0.091
64	''''C–O–C	−0.284	0.040
65	C(/)–O–C	−0.910	0.163
66	''''C–O–C\	−0.396	0.067
67	''''C–O–C(//)\	0.549	0.077
68	''''C–O–C(//)	0.491	0.317
69	''''C–O–C''''	0.640	0.222
70	−C(/)–O–C\	−0.636	0.147
71	=C(/)–O–C(//)	0.813	0.332
72	C–O–C	−1.162	0.431
73	=C–O–C(//)	0.400	0.444
74	=C(/)–O–C(/)\	0.589	0.162
75	=C(/)–N=C\	−0.526	0.158
76	=C(/)–N–C(//)	0.135	0.130
77	''''C–N–C(//)	0.118	0.104
78	''''C–N–C(//)\	0.230	0.060
79	''''C–N=C\	−0.710	0.151

CI No.	Atom with Environment	Mean Value[a,b]	Std. Dev.[b]
80	""C–N–C	−0.336	0.162
81	""C–N=Z	−0.199	0.163
82	Z–N=Z	0.110	0.172
83	Z–N=C	−0.197	0.103
84	C–N""C	−0.801	0.051
85	=C–N=C	−0.630	0.316
86	=C–N–C∥	−0.105	0.356
87	Z–N=C	−0.172	0.112
88	Z–N–C∥	−0.538	0.448
89	C–N–C	−1.665	0.219
90	=C–N–C	−0.170	0.217
91	C–N–C	−1.521	0.103
92	–C–N""C	−0.567	0.043
93	–C–N–C	−1.433	0.221
94	Z–N–C"	0.064	0.096
95	–C–N""C	−0.520	0.056
96	""C–N–C	−0.497	0.155
97	""C–N""C	−0.552	0.064

CI No.	Atom with Environment	Mean Value[a,b]	Std. Dev.[b]
98		− 0.073	0.123
99		− 0.626	0.145
100		− 0.559	0.072
101		− 0.481	0.175
102	Z=N=Z	—	—
103		0.162	0.158
104		− 0.313	0.075
105		− 0.157	0.314
106		− 0.931	0.162
107		0.970	0.072
108		1.727	0.257
109		0.016	0.131
110		− 0.407	0.132
111		0.465	0.688
112		− 0.525	0.439
113		0.091	0.162
114		0.052	0.209
115		− 0.442	0.174

CI No.	Atom with Environment	Mean Value[a,b]	Std. Dev.[b]
116	=C(O–C)	− 0.170	0.304
117	Z(O–C)	− 0.378	0.159
118	Z(O–C)	1.086	0.206
119	Z(O–C)	− 0.588	0.448
120	Z(O–C)	− 0.473	0.150
121	Z=N=C	—	—
122	C(O–C)	− 0.310	0.254
123	=C(N–C)	0.078	0.435
124	Z(N–C)	− 0.389	0.157
125	C(N–C)	0.973	0.467
126	=C(S–C)	1.500	0.139
127	C(N=C)	− 1.118	0.458
128	C(S–C)	− 0.308	0.263
129	C(S–C)	0.999	0.195
130	=C(S–C)	0.788	0.175
131	=C(S–C)	0.866	0.283
132	Z(N–C)	− 0.517	0.195
133	C(S–C)	0.405	0.130

CI No.	Atom with Environment	Mean Value[a,b]	Std. Dev.[b]
134		− 0.248	0.306
135		1.250	0.601
136		− 0.270	0.150
137		− 0.355	0.203
138		0.882	0.205
139		− 0.700	0.360
140		—	—
141		− 0.098	0.433
142		1.455	0.256
143		—	—
144		0.742	0.216
145		0.178	0.434
146		− 0.198	0.325
147		0.419	0.581
148		0.576	0.435
149		0.684	0.356
150		—	—
151		1.144	0.310

CI No.	Atom with Environment	Mean Value[a,b]	Std. Dev.[b]
152	C—S—C	—	—
153	Z—N—Z	−0.258	0.482
154	C, Z—N—C	−1.579	0.193
155	C, Z—N—C	−0.994	0.336
156	Z, Z—N—C	−0.469	0.244
157	Z, Z—N—C	0.151	0.366
158	C, C—N—C	−1.678	0.308
159	C, C—N—C	−0.194	0.333
160	C, —C—N—C	−1.621	0.117
161	C, —C—N—C	−1.984	0.112
162	C, Z—N—C	−0.714	0.566
163	C, Z—N—C	−0.657	0.260
164	Z, Z—N—C	−0.060	0.474
165	C, C—N—C	−0.212	0.079

CI No.	Atom with Environment	Mean Value[a,b]	Std. Dev.[b]
166	C, Z–N–C	− 0.795	0.169
167	C, –C–N–C	− 1.736	0.311
168	C, C–N–C	− 0.942	0.173
169	–C, =C–N–C	− 1.566	0.472
170	=C, C–N–C=	− 0.484	0.445
171	=C, C–N–C	− 0.680	0.121
172	C, C–N–C	0.775	0.437
173	C, C–N–C	− 0.781	0.433
174	C, –C–N–C=	− 0.651	0.211
175	=C, C–N–C	− 0.756	0.276
176	C, Z–N=Z	0.284	0.344
177	=C, C–N–C=	0.467	0.256

CI No.	Atom with Environment	Mean Value[a,b]	Std. Dev.[b]
178	=C, N—C, —C≡	0.790	0.140
179	C, N—C, —C≡	−0.835	0.086
180	—C, N—C, —C≡	−0.922	0.256
181	C, N, C, Z	−1.160	0.235
182	=C, N=Z, Z	−0.233	0.361
183	=C, N—C, Z	0.237	0.585
184	C, I=Z, Z	−3.157	0.550
185	C, N—C, —C≡	−0.641	0.134
186	=C, N=Z, Z	0.544	0.388
187	=C, N—C, —C≡	1.520	0.236
188	C, N—C, —C	−3.346	0.203
189	C, N=C, Z	−1.128	0.343

CI No.	Atom with Environment	Mean Value[a,b]	Std. Dev.[b]
190	C, N–C, Z	−1.382	0.101
191	=C, N–C, –C	−0.167	0.380
192	C, N–C, Z	−0.646	0.133
193	=C, N–C, C	−0.619	0.444
194	C, N=Z, Z	−0.240	0.380
195	=C, N–C, –C	0.599	0.122
196	C, N–C, C	−1.695	0.444
197	C, N=Z, Z	−0.623	0.464
198	Z, N=Z, Z	0.999	0.574
199	C, N–Z, Z	−1.953	0.369
200	=C, N–C, –C	−0.687	0.275
201	C, N–C, Z	−1.092	0.484
202	=C, N–C, Z	0.342	0.220

CI No.	Atom with Environment	Mean Value[a,b]	Std. Dev.[b]
203		−0.577	0.463
204		−1.311	0.442
205		−1.843	0.535
206		−1.098	0.251
207		−2.072	0.558
208		2.242	0.486
209		0.296	0.561
210		−2.951	0.510
211		−2.129	0.511
212		−1.608	0.550
213		−1.571	0.440
214		−2.332	0.567

CI No.	Atom with Environment	Mean Value[a,b]	Std. Dev.[b]
215	Z—P=Z with C and Z substituents	− 1.756	0.501
216	C—S=Z with Z and C substituents	− 1.588	0.696
217	Z—S=Z with Z and C substituents	− 1.975	0.578
218	Z—S=Z with Z and Z substituents	− 1.674	0.655
219	Z—P=Z with C and Z substituents	− 1.210	0.658
220	Z—P=Z with =C and Z substituents	− 0.679	0.643
221	Z—P=Z with =C and Z substituents	− 0.837	0.628
222	Conjugation	0.270	0.019

- data uncertain

[a]Obtained by regression analysis

Note: Contribution 222 is the conjugation contribution.

Source: Broto, P., G. Moreau, and C. Vandycke, Molecular Structure: Perception, Autocorrelation Descriptor and SAR Studies. System of Atomic Contributions for the Calculation of the *n*-Octanol/Water Partition Coefficients. *Eur. J. Med. Chem. Chim. Ther.*, 1984: **19**, 71–78.

APPENDIX G

GLOSSARIES

G.1 PROPERTY AND PHYSICAL STATE NOTATIONS

AWPC	air–water partition coefficient
BCF	bioconcentration factor
C_{air}	molar concentration in air
C_w	mass-per-volume concentration in water at saturation where indicated
C_w^s	water solubility expressed as molar saturation concentration
d_{t1}^{t2}	Specific gravity defined as the ratio of the weight of any volume of a given liquid at temperature t_1 to the weight of an equal volume of same standard at temperature t_2
H_c	Henry's law constant: $H_c = K_{aw} \cdot RT$
H^g	vapor phase enthalpy
H^l	liquid phase enthalpy
HLC	Henry's law constant
H_x	Henry's law constant: $H_x = H_y p_G$ where p_G = gas-phase pressure
H_y	Henry's law constant: $H_y = X_{i\,in\,air}/X_{i\,in\,water}$ (mole fraction ratio for compound i)
I	ionic strength
I_{vc}	viscosity-constitutional constant [1]
K_{aw}	air–water partition coefficient (C_{air}/C_w)
K_d	soil–water distribution coefficient
K_{oc}	soil–organic carbon partition coefficient
K_{ow}	n-octanol–water partition coefficient
K_p	Freundlich–adsorption coefficient
$K_{soil/water}$	soil–water partition coefficient
m_p	viscosity–density constant [1]
n_D^t	refractive index at temperature t measured with light of the sodium D line
$1/N$	Freundlich exponent
parachor	$= s_1^{1/4} M (r_1 - r_v)^{-1}$
P_c	critical pressure
pK_a	negative logarithm (base 10) of acid dissociation constant

p_G	total (atmospheric) pressure
p_i	partial pressure of compound i
p_v	vapor pressure
R_D^t	molar refraction at temperature t (sodium D line)
S_w	Solubility in water (mass-per-mass solubility, usually in g/g%), compare with C_w
t, T	temperature
T_b	normal boiling point ($p = 1$ atm)
T_b^p	boiling point at pressure p
T_c	critical temperature
T_m	normal melting point ($p = 1$ atm)
T_{min}	temperature of solubility minimum
T_x	temperature of interest
V_b	molar volume at normal boiling point
V_c	critical volume
V_M	molar volume
V_i	intrinsic molecular volume
V_M^t	molar volume at temperature t
X_k	mole fraction of a component k
X_k^s	mole fraction solubility (superscript s = saturation) of organic compound k in water
X_w^s	mole fraction solubility (superscript s = saturation) of water in organic compound
XYZ	solvent-dependent property, used in LSERs to indicate the dependent variable
γ^∞	activity coefficient at infinite solution
ΔG_v	Gibbs free energy change for vaporization process
ΔH_{soln}	enthalpy of solution
ΔH_v	enthalpy of vaporization
ΔH_{vb}	enthalpy of vaporization at the normal boiling point
ΔS_{fus}	entropy of fusion
ΔS_{vb}	entropy of vaporization at the normal boiling point
ΔZ	compressibility factor
η_l	liquid dynamic viscosity
η_{lb}	liquid dynamic viscosity at normal boiling point
Θ	Guldberg ratio T_b/T_c
λ_l	thermal conductivity of liquid
μ	dipole moment
ν_l	liquid kinematic viscosity
ρ_l	liquid density
ρ_v	vapor density
σ_l	liquid surface tension

REFERENCE

1. Sounders, M., Jr., Viscosity and Chemical Constitution. *J. Am. Chem. Soc.*, 1938: **60**, 154–158.

G. 2 MOLECULAR DESCRIPTOR NOTATIONS

Δ_{vdW}	van der Waal's surface area
CPSA	charged partial surface area [1]

CRI   characteristic root index [2]

DEX   descriptor of existence [3]

EAmax   extended-adjacency-matrix descriptor [4]

EAS   extended-adjacency-matrix descriptor [4]

FNSA   fractional negative surface area [1]

FPSA   fractional positive surface [1]

FREL   fragment reduced to an environment that is limited [3]

$FREL_B$ atom-centered substructure used with the LOGIC method [5]

GAI   general a_N index [6]

GSA   group surface area [7]

ICI   information content index

I_{mean} mean information content (information theoretical index) [8]

I_{total} total information content

J Balaban index

L / B   length to breath of rectangle enclosing planar molecule [8,9]

M molar mass (formerly known as molecular weight)

MTI   molecular topological index [10]

N_{Br} number of bromine atoms per molecule

N_C number of carbon atoms per molecule

N_{CF2} number of perfluoromethylene groups per molecule

N_{CH2} number of methylene groups per molecule

N_{Cl} number of chlorine atoms per molecule

N_{Cq} number of quaternary carbon atoms per molecule

N_{Ct} number of tertiary carbon atoms per molecule

N_F number of fluorine atoms per molecule

N_H number of hydrogen atoms per molecule

N_I number of iodine atoms per molecule

N_N number of nitrogen atoms per molecule

N_{NH} number of imino groups per molecule

N_{NH2} number of amino groups per molecule

N_{NO2} number of nitro groups per molecule

N_O number of oxygen atoms per molecule

N_{OH_s} number of hydroxyl groups attached to a secondary carbon atom

N_{OH_t} number of hydroxyl groups attached to a tertiary carbon atom

N_{ortho} number of substituent pairs in ortho position at benzene ring

N_P number of phosphorous atoms per molecule

N_S number of sulfur atoms per molecule

$N_=$ number of double bonds per molecule

$N_\#$ number of triple bonds per molecule

P_4' number of pairs of quaternary C atoms separated by one C atom [11]

P_4'' number of pairs consisting of one quaternary and one tertiary C atom separated by one other C atom [11]

PCdC   number of pairs of atoms d bonds apart [11]

P_{cdo} number of pairs consisting of one C and one O atom that are d bonds apart [11]

PNSA   partial negative surface area [1]

PPSA   partial positive surface area [1]

RNSA   relative negative surface area [1]

RPSA   relative positive surface area [1]

T_3' number of terminal methyl groups separated by three bonds

TSA   total surface area

V_g geometric volume of molecule [12]

V_H "hydrophilic effect" volume [13]

V_I	Intrinsic molecular volume [14]
V_L	$= V_{vdW} - V_H$ [13]
V_{vdW}	van der Waals volume
W	Wiener index
WNSA	weighted negative surface area [1]
WPSA	weighted positive surface area [1]
α	electric polarizability [15]
ΔP_{C3C}	$= P_{C3C}$ (branched isomer)-P_{C3C}(normal isomer) [11]
ε	"edge-adjacency" index [16]
σ_{sym}	rotational symmetry number [17]
Φ	polarizability factor [18]
ϕ_{flx}	conformational flexibility number [17]
χ	Randic branching index; extension of the Randic approach by Kier and Hall [19]

REFERENCES

1. Stanton, D. T., and P. C. Jurs, Development and Use of Charged Partial Surface Area Structural Descriptors for Quantitative Structure–Property Relationship Studies. *Anal. Chem.*, 1990: **62**, 2323–2329.

2. Saçan, M. T., and Y. Inel, Application of the Characteristic Root Index Model to the Estimation of *N*-Octanol / Water Partition Coefficients: Polychlorinated Biphenyls. *Chemosphere*, 1995: **30**, 39–50.

3. Carrier, G., A. Panaye, an J.-É. Dubois., Topological Structural Information in the CAS File: Statistical Occurrences of DARC Concentric Fragments: 1. Basic Carbon Substructures. *J. Chem. Inf. Comput. Sci.*, 1990: 30, 110–121.

4. Yang, Y.-Q., L. Xu, and C.-Y. Hu, Extended Adjacency Matrix Indices and Their Applications. *J. Chem. Inf. Comput. Sci.*, 1994: **34**, 1040–1145.

5. Dubois, J. E., and M. Loukianoff, DARC "Logic Method" for Molal Volume Prediction. *SAR QSAR Environ. Res.*, 1993: **1**,d 63–75.

6. Xu, L., H.-Y. Wang, and Q. Su, A Newly Proposed Molecular Topological Index for the Discrimination of Cis / Trans Isomers and for the Studies of QSAR / QSPR. *Comput. Chem.*, 1992: **16**, 187–194.

7. Amidon, G. L., and S. T. Anik, Application of the Surface Area Approach to the Correlation and Estimation of Aqueous Solubility and Vapor Pressure: Alkyl Aromatic Hydrocarbons. *J. Chem. Eng. Data*, 1981: **26**, 28–33.

8. Wise, A. A., et al., A Relationship Between Reversed-Phase C18 Liquid Chromatographic Retention and the Shape of Polycyclic Aromatic Hydrocarbons. *J. Chromatogr. Sci.*, 1981: **19**, 457–465.

9. Jinno, K., and K. Kawasaki, Correlation Between the Retention Data of Polycyclic Aromatic Hydrocarbons and Several Descriptors in Reversed-Phase HPLC. *Chromatographia*, 1983: **17**, 445–449.

10. Schultz, H. P., Topological Organic Chemistry: 1. Graph Theory and Topological Indices of Alkanes. *J. Chem. Inf. Comput. Sci.*, 1989: **29**, 227–228.

11. Greenshields, J. B., and F. D. Rossini, Molecular Structure and Properties of Hydrocarbons and Related Compounds. *J. Phys. Chem.*, 1958: **62**, 271–280.

12. Bhattacharjee, S., and P. Dasgupta, Molecules Property Correlation in Alkanes with Geometric Volume. *Comput. Chem.*, 1994: **18**, 61–71.

13. Moriguchi, I., Y. Kanada, and K. Komatsu, Van der Waals Volume and the Related Parameters for hydrophobicity in Structure–Activity Studies. *Chem. Pharm. Bull.*, 1976: **24**, 1799–1806.

14. Leahy, D. E., Intrinsic Molecular Volume as a Measure of the Cavity Term in linear Solvation Energy Relationships: Octanol–Water Partition Coefficients and Aqueous Solubilities. *J. Pharm. Sci.*, 1986: **75**, 629–636.

15. Lewis, D. F. V., The Calculation of Molar Polarizabilities by the CNDO/2 Method: Correlation with the Hydrophobic Parameter, Log *P. J. Comput. Chem.*, 1989: **10**, 145–151.

16. Estrada, E., Edge Adjacency Relationships and a Novel Topological Index Related to Molecular Volume. *J. Chem. Inf. Comput. Sci.*, 1995: **35**, 31–33.

17. Nirmalakhandan, N. N., and R. E. Speece, Prediction of Aqueous Solubility of Organic Chemicals Based on Molecular Sturucture: 2. Application to PNAs, PCBs, PCDDs, etc. *Environ. Sci. Technol.*, 1989: **23**, 708–713.

18. Mishra, D. S., and S. H. Yalkowsky, Estimation of Vapor Pressure of Some Organic Compounds. *Ind. Eng. Chem. Res.*, 1991: **30**, 1609–1612.

19. Kier, L. B., and L. H. Hall, *Molecular Connectivity in Chemistry and Drug Research*, 1976. San Diego, CA: Academic Press.

G. 3 COMPOUND CLASS ABBREVIATIONS

CB	chlorobenzenes (mono-and polychlorinated benzenes)
DNPH	dinitrophenanthrenes
HC	hydrocarbons
HCFC	hydrochlorofluorocarbons
HFC	hydrofluorinated hydrocarbons
PAH	polycyclic aromatic hydrocarbons
PCB	polychlorinated bipenyles
PCDD	polychlorinated dibenzo-*p*-dioxines
PCDE	polychlorinated diphenyl ethers
PCDF	dibenzofuranes
TCBT	tetrachlorobenzyltoluenes
THM	trihalomethanes
TNPH	trinitrophenanthrenes

G. 4 ABBREVIATIONS FOR MODELS, METHODS, ALGORITHMS, AND RELATED TERMS

A	adjacency matrix
AMP	Abrams–Massaldi–Prausnitz model [1]
AQUAFAC	aqueous solubility model [2,3]
CANGEN	canonicalization of structure and generation of unique SMILES notation [4–6]
CASE	computer-aided structure evaluation (e.g., [7])
CLOGP	K_{ow} and R_D estimation program [8]
D	distance matrix
2D	two-dimensional
3D	three-dimensional
G	molecular graph
GCM	group contribution method
GIM	group interchange method [9]
GIPF	general interaction property function [10]
HBA	hydrogen-bond acceptor

HBD	hydrogen-bond donor
LFER	linear free energy relationship
LOGIC	local-to-global-information contribution [11]
LSER	linear solvation energy relationship
MCS	maximum common subgraph
OIGLA	orbital interaction graph of linked atoms in G [12]
OIMLA	orbital interaction matrix of linked atoms (based on OIGLA)
PCKOC	K_{oc} estimation program [13]
POLLY	program to calculate molecular graph invariants [copyright the University of Minnesota, 1988]
QPPR	quantitative property–property relationship
QSPR	quantitative structure–property relationship
QSRR	quantitative structure–retention relationship
SAW	self-avoiding walks in G
SRW	self-returning walks in G
SMILES	simplified molecular input line entry system [4–6]
SmilogP	K_{ow} estimation program [14]
UNIFAC	UNIQUAC functional group activity coefficient
UNIQUAC	universal quasi-chemical [15]

REFERENCES

1. Macknick, A. B., and J. M. Prausnitz, Vapor Pressures of Heavy Liquid Hydrocarbons by a Group Contribution Method. *Ind. Eng. Chem. Fundam.*, 1979: **18**, 348–351.

2. Myrdal, P., et al., AQUAFAC 1: Aqueous Function of group Activity Coefficients: Application to Hydrocarbons. *Chemosphere*, 1992: **24**, 1047–1061.

3. Myrdal, P., et al., AQUAFAC: Aqueous Functional Group Activity Coefficients. *SAR QSAR Environ. Res.*, 1993: **1**, 53–61.

4. Weininger, D., SMILES, a Chemical Language and Information System: 1. Introduction to Methodology and Encoding Rules. *J. Chem. Inf. Comput. Sci.*, 1988: **28**, 31–36.

5. Weininger, D., A. Weininger, and J. L. Weininger, SMILES: 2. Algorithm for Generation of Unique SMILES Notation. *J. Chem. Inf. Comput. Sci.*, 1989: **29**, 97–101.

6. Weininger, D., SMILES: 3. Depiction of Graphical Depiction of Chemical Structures. *J. Chem. Inf. Comput. Sci.*, 1990: **30**, 237–243.

7. Klopman, G., and S. Wang, A Computer Automated Structure Evaluation (CASE) Approach to Calculation of Partition Coefficient. *J. Comput. Chem.*, 1991: **12**, 1025–1032.

8. Manual., S., *MedChem Manual: Release 3.*, 1985. Claremont, CA: Pomona College.

9. Drefahl, A., and M. Reinhard, Similarity-Based Search and Evaluation of Environmentally Relevant Properties for Organic Compounds in Combination with the Group Contribution Approach. *J. Chem. Inf. Comput. Sci.*, 1993: **33**, 886–895.

10. Murray, J. S., et al., Statistically-based Interaction Indices Derived from Molecular Surface Electrostatic Potentials: A General Interaction Properties Function (GIPF). *J. Mol. Struct. (Theochem.)*, 1994: **307**, 55–64.

11. Dubois, J. E., and M. Loukianoff, DARC "Logic Method" for Molal Volume Prediction. *SAR QSAR Environ. Res.*, 1993: **1**, 63–75.

12. Xu, L., H.-Y. Wang, and Q. Su, A Newly Proposed Molecular Topological Index for the Discrimination of Cis / Trans Isomers and for the Studies of QSAR / QSPR. *Comput. Chem.*, 1992: **16**, 187–194.

13. Meylan, W. M., and P. H. Howard, Soil / Sediment Adsorption Constant Program PCKOC, 1992. Boca Raton, FL: Lewis Publishers.

14. Convard, T., et al., SmilogP: A Program for a Fast Evaluation of Theoretical Log P from the Smiles Code of a Molecule. *Quant. Struct.-Act. Relat.*, 1994: **13**, 34–37.

15. Fredenslund, A., R. L. Jones, and J. M. Prausnitz, Group-Contribution-Estimation of Activity Coefficients in Nonideal Liquid Mixtures. *AIChE J.*, 1975: **21**, 1086.

INDEX